Ia Antigens

Volume II

Man and Other Species

Editors

Soldano Ferrone, M.D., Ph.D.
Professor
Department of Pathology
College of Physicians and Surgeons
Columbia University
New York, New York

Chella S. David, Ph.D.
Professor of Immunology
Mayo Medical School
Rochester, Minnesota

CRC Press
Taylor & Francis Group
Boca Raton London New York

CRC Press is an imprint of the
Taylor & Francis Group, an **informa** business

First published 1982 by CRC Press
Taylor & Francis Group
6000 Broken Sound Parkway NW, Suite 300
Boca Raton, FL 33487-2742

Reissued 2018 by CRC Press

Library of Congress Cataloging in Publication Data
Main entry under title:
Ia Antigens.

Bibliography:
Includes index.
Contents: v. 1. Mice -- v. 2. Man and other species.
1. Ia antigens. I. Ferrone, Soldano, 1940- . II. David Chella S.
QR184.4.I244 616.07'92 82-1207
ISBN 0-8493-6461-2 (v. 1) AACR2
ISBN 0-8493-6462-0 (v. 2)

A Library of Congress record exists under LC control number: 82001207

ISBN 13: 978-1-315-89422-5 (hbk)
ISBN 13: 978-1-351-07332-5 (ebk)

Visit the Taylor & Francis Web site at http://www.taylorandfrancis.com and the
CRC Press Web site at http://www.crcpress.com

PREFACE

The significant role played by the gene products of I regions in cell-cell interactions has stimulated interest in the characterization of Ia antigens. Through the efforts of several investigators much information about the functional and structural properties of Ia antigens has been accumulated in recent years. The purpose of this book is to provide the reader with a ready source of information regarding Ia antigens in several animal species. This book is not intended to cover all aspects of Ia antigens, since we have tried to avoid overlaps with other recent publications and therefore we have selected topics which had not been discussed in other recent books and monographs. Furthermore, some topics are not discussed since the invited contributors could not meet the deadline and we did not want to delay the publication of the book.

Most of the chapters can be read as free-standing articles on that particular subject. This has led to a certain unavoidable amount of duplication and to occasional areas of controversy. We hope that this will give the reader a more complete picture of the status of the art in the field and will stimulate additional investigations to resolve the conflicting data in the literature. We have encouraged the contributors to present unpublished data in their chapters so that this book serves not only as a reference book but is also of interest to those who work in the field. Extensive bibliographies are included in several chapters in order to provide the interested reader with access to greater detail than can be included in an already large book.

We thank the contributors for having managed to find time in their already overcrowded schedules to write their chapters and for having helped us to meet the deadline. We are grateful for the patient indulgence of those who put up with some editorial fussing meant to improve the accuracy of the papers. We are indebted to the staff of CRC for their help in the preparation of this book.

Soldano Ferrone
Chella David

THE EDITORS

Soldano Ferrone received his M.D. and Ph.D. degrees from the University of Milan, Italy. Currently, he is a Professor in the Department of Pathology, College of Physicians and Surgeons, Columbia University in New York. Prior to his present position, Dr. Ferrone was an Associate Member at Scripps Clinic and Research Foundation in La Jolla, California and Adjunct Professor of Pathology at the University of California at San Diego.

Dr. Ferrone has been the recipient of many honors and awards, including the National Brigatti Award for research in hematology (1965), National Piccinini Award for research in immunohematology (1966), National Luigi Mangiagalli Award for research in hematology (1967), Butturini Award for Outstanding Research in lymphocyte biology (1968), and the European Ganassini Award for research on autoimmune hemolytic anemia (1969), a Senior Fellowship from the American Cancer Society (1974), and an Established Investigatorship from the American Heart Association (1976).

Chella S. David received his undergraduate degree from Berea College, Berea, Kentucky in 1961 and his M.S. degree from the University of Kentucky in 1962. He received his Ph.D. in Immunogenetics in 1966 from Iowa State University. Currently, Dr. David is a Professor of Immunology at the Mayo Medical School in Rochester, Minnesota.

Dr. David is a member of the Genetics Society of America, the American Association of Advancement of Science, the American Association of Immunologists, and the Transplantation Society of America.

CONTRIBUTORS

N. Ade
Department of Somatic Genetics
Institute of Immunology and Genetics
Heidelberg, West Germany

E. D. Albert, M.D.
Professor of Immunogenetics
University of Munich
Munich, West Germany

Edward J. Ball
Fellow in Immunology
Department of Internal Medicine
University of Texas Health Science Center
Dallas, Texas

Hans Balner, M.D.
Professor of Immunogenetics
Director, Primate Center TNO
Rijswijk, Holland

B. Beck
Department of Immunology
Mayo Clinic and Medical School
Rochester, Minnesota

Reinhard Burger, M.D.
Research Associate
Institut f. Med. Mikrobiologie
Mainz, West Germany

Max D. Cooper, M.D.
Professor of Pediatrics and Microbiology
Director
Cellular Immunobiology Unit of the Tumor Institute
University of Alabama
Birmingham, Alabama

Chella S. David, Ph.D.
Professor of Immunology
Mayo Medical School
Rochester, Minnesota

Donald L. Ewert, Ph.D.
Assistant Professor
The Wistar Institute
Philadelphia, Pennsylvania

C. Garrison Fathman, M.D.
Associate Professor
Department of Medicine
Stanford University Medical School
Stanford, California

Marc Fellous, M.D., Ph.D.
Laboratoire d'Immuno-Hematologie
Hospital Saint-Louis
Paris, France

Soldano Ferrone, M.D.
Professor
Department of Pathology
College of Physicians and Surgeons
Columbia University
New York, New York

John H. Freed, Ph.D.
Assistant Professor
Department of Biophysics
Johns Hopkins University School of Medicine
Baltimore, Maryland

Jeffrey A. Frelinger, Ph.D.
Associate Professor
Department of Microbiology
University of Southern California Medical School
Los Angeles, California

Patrizio Giacomini
Research Fellow
Department of Pathology
College of Physicians and Surgeons
Columbia University
New York, New York

R. Grützmann, Ph.D.
Bayer, AG
Wuppestal, West Germany

G. J. Hämmerling, Ph.D.
Professor and Chairman
Department of Somatic Genetics
Institute of Immunology and Genetics
Heidelberg, West Germany

CONTRIBUTORS (continued)

Terry J. Higgins, Ph.D.
Associate Professor
Department of Microbiology
School of Medicine
University of Pennsylvania
Philadelphia, Pennsylvania

Francesco Indiveri
Associate Professor
Medical School, I.S.M.I.
University of Genova
Genova, Italy

Jorge E Kalil, M.D.
Laboratoire d'Immuno-Hematologie
Hospital Saint-Louis
Paris, France

Masao Kimoto, M.D.
The Third Department of Internal Medicine
Osaka University Hospital
Osaka, Japan

N. Koch, Ph.D.
Department of Somatic Genetics
Institute of Immunology and Genetics
Heidelberg, West Germany

William P. Lafuse
Assistant Professor
Department of Immunology
Mayo Medical School
Rochester, Minnesota

Ian F. C. McKenzie, M.D.
Professor
Department of Pathology
Melbourne University
Parkville, Victoria
Australia

Pier Giorgio Natali
Assistant Professor
Instituto Regina Elena
Roma, Italy

Ah-Kau Ng
Assistant Professor
Department of Pathology
College of Physicians and Surgeons
Columbia University
New York, New York

Gabriel Nunez
Fellow in Immunology
Department of Internal Medicine
University of Texas Health Science Center
Dallas, Texas

Jun Okada
Fellow in Immunology
Department of Internal Medicine
University of Texas Health Science Center
Dallas, Texas

Christopher R. Parish, Ph.D.
Senior Fellow
Department of Microbiology
John Curtin School of Medical Research
Australian National University
Canberra, Australia

Michele A. Pellegrino, M.D.
Associate Professor
Department of Pathology
College of Physicians and Surgeons
Columbia University
New York, New York

Carlo Russo
Instructor
Department of Pathology
College of Physicians and Surgeons
Columbia University
New York, New York

David H. Sachs, M.D.
Chief, Transplantation Biology Section
Immunology Branch
National Cancer Institute
National Institutes of Health
Bethesda, Maryland

CONTRIBUTORS (continued)

Ronald H. Schwartz, M.D., Ph.D.
Senior Investigator
Laboratory of Immunology
National Institute of Allergy and Infectious Diseases
National Institutes of Health
Bethesda, Maryland

Ethan M. Shevach, M.D.
Senior Investigator
Laboratory of Immunology
National Institute of Allergy and Infectious Diseases
National Institutes of Health
Bethesda, Maryland

M. Shigeta
Department of Medicine
Stanford University Medical School
Stanford, California

Nobukata Shinohara, M.D., Ph.D.
Associate Professor of Medicine
Department of Immunology
Chiba University School of Medicine
Chiba City, Japan

Peter Stastny, M.D.
Professor of Internal Medicine
University of Texas Health Science Center
Dallas, Texas

Ia Antigens

Soldano Ferrone and Chella S. David

Volume I: Mice

Volume II: Man and Other Species

TABLE OF CONTENTS

Volume II

Chapter 1

Ia-LIKE ANTIGENS OF THE CHICKEN

Donald L. Ewert and Max D. Cooper

TABLE OF CONTENTS

I. INTRODUCTION

The B complex of genes[1,2] in the chicken is considered analogous to the major histocompatibility complex (MHC) of mammals (H-2 of mice or HLA of man). Based on genetic and immunochemical evidence,[3] the antigenic products of three subregions of the B complex can be distinguished: (1) the B-F subregion which controls expression of antigens that are similar in molecular weight, to the K and D molecules of mice (40,000–43,000 mol wt), that are associated with a β-2 microglobulin-like molecule (12,000 mol wt) and that occur on both lymphocytes and erythrocytes;[4] (2) the B-G subregion, which determines expression of 43,000 mol wt antigens found only on erythrocytes and which have no known mammalian counterparts;[7] and (3) the B-L subregion, which controls expression of antigens that structurally resemble the Ia antigens of mammals, are not associated with a β-2 microglobulin-like molecule and are found on white blood cells but not erythrocytes. Among the three known MHC regions, only one recombination event, that between the B-F and B-G subregions, has been documented.[5]

II. GENETICS

Two lines of evidence exist for the location of genes within the B complex which encode for Ia-like molecules. In both cases the determination was dependent on the availability of genetically defined lines of chicken. Zeigler and Pink[6,7] raised antisera to B region blood group antigens by cross immunization with whole blood between two highly inbred lines of chickens, CC (genotype B_1/B_1) and CB (B_2/B_2). These two lines were previously described[8,9] as congenic lines with different MHCs on otherwise identical genetic backgrounds. Consequently, any antigenic differences detected by alloimmunization were presumed to be controlled by the B region. One antiserum (anti-B_2) precipitated a 30,000 mol wt antigen that was shown by absorption analysis to be present on peripheral white blood cells but not erythrocytes, and by immunofluorescence to stain selectively immunoglobulin positive (Ig^+) peripheral blood lymphocytes. Two-dimensional electrophoresis was used to resolve the single band into two groups of spots.[10]

Antisera with similar specificity were obtained in our laboratory using highly inbred lines of chickens that were not congenic.[11] We raised alloantisera by reciprocal immunization with lymphocytes in lines 7_2 and $15I_5$ chickens, both of which had inbreeding coefficients greater than 0.99[11] and were homozygous for the B^2 and B^{15} erythrocyte alloalleles of the MHC, respectively. Some of the resulting antisera reacted exclusively with Ig^+ lymphocytes and a subpopulation of Ig^- monocytes without prior absorption. Other antisera had to be absorbed with erythrocytes to remove antibody reactive with MHC antigens expressed on all nucleated cells. Proof that the alloantigens detected by those antisera were not allotypic determinants of immunoglobulins on B cells was shown by their antibody-mediated redistribution independent of immunoglobulin determinants. Based on their cellular distribution, these antigens were designated Ia-like. Both Ia alleles were found to be codominantly expressed on lymphocytes of F1 hybrid chickens. Formal genetic analysis by backcrossing ($15I_5 \times 7_2$) hybrids to chickens from either of the parental lines failed to produce a single recombinant between genes coding for B (MHC) erythrocyte antigens and Ia-like antigens among 208 progeny.[12] These results demonstrate that the genes coding for the Ia-like antigens are a part of the B complex.

Anti-Ia antisera produced in noncongenic lines of chickens are useful for immunofluorescence and cytotoxic assays, but have generally been ineffective for immunoprecipitation of solubilized antigens. Presumably this is a consequence of the low titer and affinity of the antibodies produced by this regimen. Continual immunization of inbred chickens, in which the initial antibody response is directed against the Ia alloantigens of the donor cells, results in stimulation of antibodies to other alloantigens. Thus the production of high titer and high affinity antibody with narrow specificity comparable to those obtained in congenic mice which differ only at subregions of the H-2 complex has not been possible in chickens. To circumvent this problem, we have prepared mouse monoclonal antibodies to avian Ia antigens using splenic lymphocytes for immunization.[13] One such antibody, an IgM, kappa mouse monoclonal antibody, was shown by immunofluorescence to co-cap cell membrane antigens stained by alloanti-Ia antibody and to co-stain the same cell populations as the alloantibody. The monoclonal antibody apparently binds a constant, nonallotypic portion of the Ia molecule as indicated by its reaction with B lymphocytes from all avian species tested, i.e., chicken, duck, quail, dove, and pigeon.[9] This antibody has been useful for the structural analysis of the Ia molecules.

III. FUNCTIONAL RELATIONSHIPS

An important consideration in the designation of antigens as Ia antigens is their association with genes that control immunologic functions. In both mice[14] and humans,[15,16] genes controlling expression of Ia antigens or DR antigens, respectively, are associated with in vitro stimulation of T lymphocyte proliferation in mixed lymphocyte assays. Genes within the B complex of chickens are known to affect levels of stimulation obtained in a mixed lymphocyte response (MLR).[17] Of the three known regions of the B complex, only the B-F and B-L regions control expression of antigens on lymphocytes. The observation of Jones[18] that erythrocytes do not significantly stimulate in an MLR diminishes the likelihood that the B-F region effects stimulation. In addition, Simonsen[19] found that two inbred lines of chickens having serologically similar Ia (B-L encoded) antigens but markedly different B-F antigens, did not stimulate each other's cells as well as do third party haplotypes in an MLR. This suggests that the B-L region and not the B-F region controls MLRs. Furthermore, we have found that mouse monoclonal anti-Ia antibody completely blocks the MLR.[13] In this experiment, we were not able to determine whether the stimulator or responder cells were affected by the anti-Ia antibody, since the antibody was directed against nonallelic determinants of the Ia antigen. We have observed, as have others, that alloanti-Ia antisera do not significantly reduce the appropriate MLRs, most likely because of the low titer and affinity of the specific antibody. Thus, while the evidence suggests that this Ia antigen of chickens plays a role in lymphocyte stimulation similar to that of mamalian Ia or DR antigens, critical blocking experiments with alloantibody have not been reported.

Another observation that indicates an immunoregulatory role of B region genes was made by Toivanen and Toivanen[20] using cyclophosphamide-treated chicks reconstituted with combinations of histocompatible, semiallogeneic or allogeneic B and T cells. They found that germinal center formation was dependent on serologically detectable similarities in the B region-controlled antigens of the two cell populations. These findings are similar to the Ia allotype-dependent recognition between B cells, T cells, and macrophages required for eliciting immune responses in mice[21] and guinea pigs.[22]

Furthermore, it is well established in fowl, as in several mammalian species, that certain genes (Ir genes) within the MHC control immune responses to conventional antigens. Associations between serologically determined MHC antigens on erythrocytes in chickens and immune responses have been documented for *Salmonella pullorum* bacteria,[23-25] the dinitrophenol group conjugated to chicken IgG,[26] the synthetic polypeptides (T,G)-A--L,[27,28] GAT,[29] and tuberculin.[30] Also, Pevzner et al.[31] have documented a crossover between the Ir region controlling the immune response to GAT and the genes (B-F) that determine the MHC antigens found on erythrocytes, indicating that the two regions are linked but not identical. Therefore, proof for the association of the B-L region encoded Ia antigens and the Ir region of the chicken is at best circumstantial in that both regions are closely linked to the genes (B-F) controlling expression of B erythrocyte antigens. It would be of interest to determine if the high and low response to GAT in the chickens used by Pevzner et al.[31] are linked to distinct Ia alleles.

IV. STRUCTURE

Structurally the Ia antigens of chickens are remarkably similar to their mammalian counterparts. Using the techniques of endogenous amino acid radiolabeling, immune precipitation, and SDS-polyacrylamide gel electrophoresis (SDS-PAGE), the Ia antigens can be resolved into two polypeptide chains with molecular weights of approximately 32,000 and 27,000 corresponding to the α and β chains, respectively, of mammalian Ia molecules (Figure 1a). Also, the observation that the smaller (β) chain is more polymorphic than the α chain when analyzed by two-dimensional gel electrophoresis,[10] is similar to findings for murine Ia and human DR antigens.[23,33] Pink,[10] using nonequilibrium pH gradient electrophoresis, has resolved a third extra basic polypeptide that may correspond to the invariant I chain described by Jones et al.[34] for mouse Ia antigens.

The native association of the α and β chains does not involve disulfide linkages, since they exist as separate units under nonreducing conditions. However, a 65,000 mol wt species can be obtained if immune complexes of the Ia antigen are maintained at room temperature in SDS sample buffer rather than boiled in the same buffer prior to PAGE analysis (Figure 1b). The association of the Ia chains does not seem to involve the artificially induced disulfide linkages that have been reported to be formed during solubilization of murine Ia molecules,[35,36] as iodoacetamide (which prevents disulfide bonding) was present in the lysing medium. Pink[37] has also observed slower moving peaks under reducing conditions, suggesting the presence of an intrachain disulfide bond which results in a more compact conformation of the nonreduced molecule.

Based on the above evidence, the avian Ia antigens examined thus far resemble the mouse I-E/C or human DR antigens more closely than the murine I-A antigens. The α and β chains of mouse I-A antigens are not dissociated under nonreducing conditions, whereas under the same conditions, both the I-E/C and DR antigens are dissociated.[14,15] Also, the relative mobility of α and β chains of the I-E/C and DR antigens are affected by reduction,[15] but the I-A antigen chains are not.

The association of carbohydrate moieties with either or both of the avian Ia antigen chains is indicated by their binding by lectin[10] and subsequent elution with sugars.[13] Brogren et al.[38] have found that *Lens culinaris* lectin will bind avian B (B-F) antigens, immunoglobulin, and Ia antigens, whereas *Ricinus communis* lectin binds only Ia antigens and immunoglobulin, thus providing a means of separating the Ia and B (B-F) antigens. This evidence, coupled with the demonstration of tritium-labeled amino acid incorporation by both chains of the Ia antigen,[10,38] indicates that these antigens are glycoproteins. However, the chemical nature of the allotypic determinants is not known for certain. Incubation of intact cells with trypsin, pronase, or papain under conditions that remove surface IgM did not prevent detection of the antigens with alloanti-Ia antisera.[11] Also, treatment with neuraminidase, which removes sialic acid residues, had no apparent effect on staining of either cell surface IgM or Ia. Since the conditions for enzyme treatment were limited in order to preserve cellular integrity, the results may reflect only the relative sensitivities of IgM and Ia antigenic determinants rather than their absolute chemical composition.

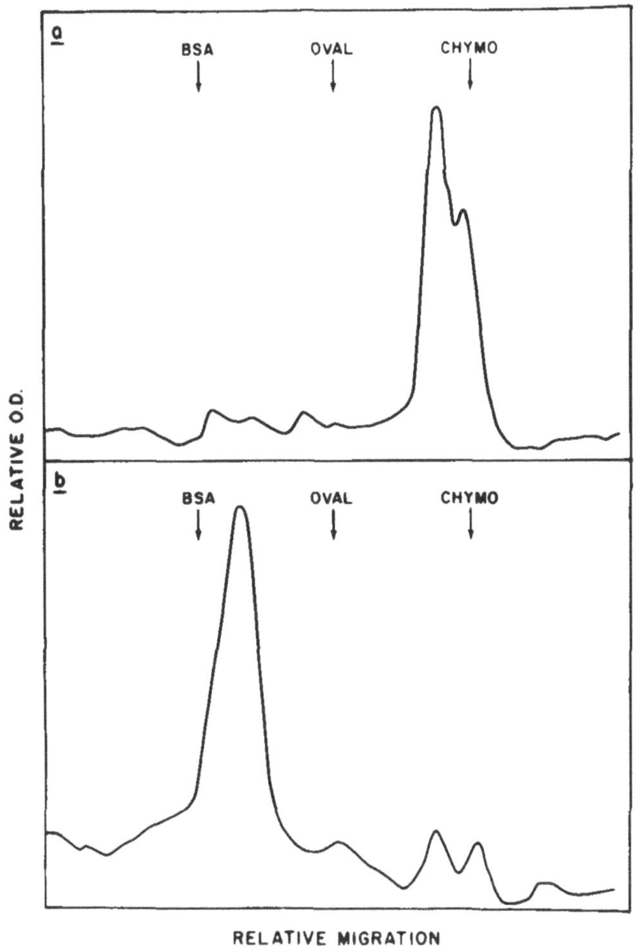

FIGURE 1. SDS-PAGE analysis of immunoprecipitates of monoclonal anti-Ia antibody and NP-40 solubilized membranes of ^{125}I lactoperoxidase-labeled splenic leukocytes: (1a) samples were boiled in SDS buffer containing 2-mercaptoethanol; (1b) samples contained 0.01 *M* iodoacetamide and were held at room temperature or below prior to electrophoresis. Standards run in adjacent wells of a 5 to 15% gradient Laemli slab gel were BSA (65K), ovalbumin (45K), and chymotrypsinogen (25K). Ordinate represents OD_{470} of microdensitometer tracing of gel autoradiograph.

Structural similarity of avian and mammalian Ia antigens is also indicated by the finding that rabbit antisera against HLA-DR antigens will precipitate analogous 2-chain molecules from human, mouse, and chicken lymphocytes (see Figure 2) as well as from monkey, rat, guinea pig, cow, and pig lymphocytes.[39] However, mouse monoclonal anti-chicken-Ia antibody does not stain lymphocytes of mouse, man, reptiles, or fish, but does react with B lymphocytes of all avian species tested.[13]

V. TISSUE DISTRIBUTION

In the chicken, Ia antigens have been demonstrated by immunofluorescence on B cells, macrophages, and an unidentified subpopulation of mononuclear cells in bone marrow, spleen, thymus, and bursa of Fabricius[11] (Figure 3). Thrombocytes, thymocytes, erythrocytes, fibroblasts, spermatozoa, kidney cells, embryonic liver cells, and

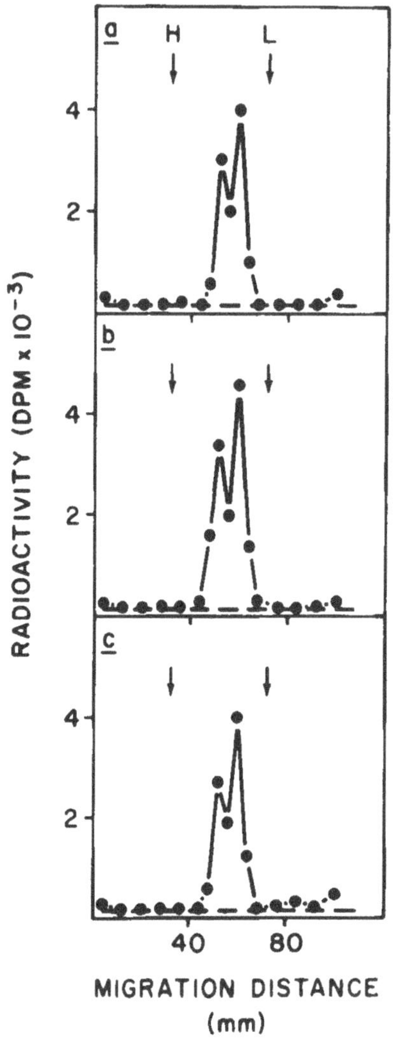

FIGURE 2. SDS-polyacrylamide gel electrophoresis of ^{3}H-tyrosine-labeled membrane glycoproteins precipitated with a rabbit anti-HLA-DR antigen serum (○----○) and normal rabbit serum (●----●), respectively, from human (a), mouse (b), or chicken (c) lymphocytes. The arrows mark the migration positions of IgG heavy (H) and light (L) chains. (From Kvist, et al., *Scand. J. Immunol.*, 7, 477, 1978. With permission.)

embryonic yolk sac cells were not stained. In addition to B cells of normal tissues, all lymphoid leukosis tumor cells examined express Ia antigens as well as surface IgM. Ia antigens have not been detected on normal unstimulated T cells or tumors of T cell origin.[11,40]

The level of Ia expression, as indicated by the intensity of immunofluorescence, varies widely among cells of the B cell lineage. The intensity of staining for Ia determinants appeared uniformly bright for all Ig^{+} lymphocytes in peripheral blood, spleen,

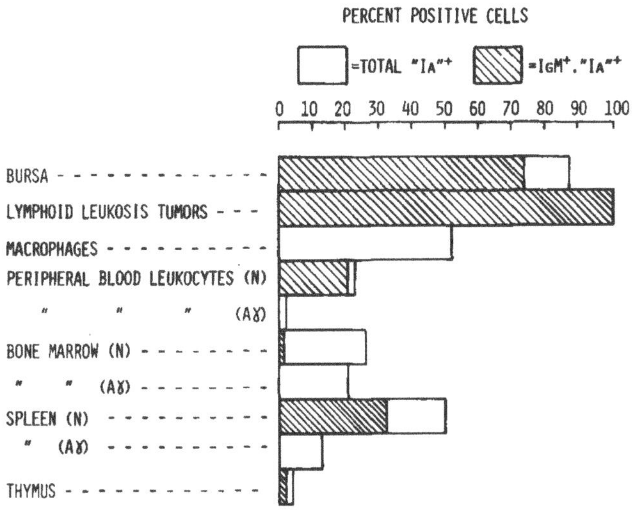

FIGURE 3. Distribution of surface Ia$^+$ and IgM$^+$ cells in different lymphoid and hemopoietic tissues in normal (N) and agammaglobulinemic (Aγ) chickens. all values represent mean of two or more determinations. Cells were stained with fluorochrome-labeled goat anti-μ chain antibody and indirectly with alloanti-Ia antibody followed by goat anti-γ chain conjugated with a contrasting fluorochrome. IgG$^+$ cells represented 0.01% 0.1%, 1 to 3%, and 4 to 7% of the cells in bone marrow, bursa, and thymus, blood leukocytes and spleen, respectively.

thymus, and bone marrow, whereas bursal cell and plasma cell staining was variable.[11] Various assays and investigators have given different percentages of Ia$^+$ cells in the bursa, e.g., immunofluorescence gave 73 to 88%,[11] 56 to 74%,[41] or about 50%;[42] cytotoxicity gave 73%,[13] and rosette assay gave 48 to 71%.[41] These variations may be due to differences in antibody titer as well as to the sensitivity of the assay. We found that the percentage of bursal cells staining indirectly with Ia antisera was dramatically reduced upon dilution of the alloantisera.[11] We also observed that direct staining, with fluorochrome-labeled monoclonal anti-Ia antibody, stains primarily large cells which represent 40 to 60% of the bursal population. When the sensitivity of the assay is increased by indirect staining, the same monoclonal antibody stains an additional 20 to 30% of the bursal cells. These differences may reflect variable expression of the Ia antigens during the cell cycle or different levels of maturation of the bursal cells.

Although the expression of Ia antigens extends to the plasma cell level of B cell differentiation, only a fraction of the plasma cells have sufficient antigen to permit detection by immunofluorescence. Using mouse monoclonal anti-Ia antibody, the percentage of plasma cells expressing cytoplasmic IgM, IgG, or IgA that bore surface Ia antigens was 30%, 12% and 9%, respectively.[13] The significance of these observations is not entirely clear. In general, the concentration of Ia antigens on cells of the B cell lineage appears to increase as cells mature and then decrease as the terminal, plasma cell stage of differentiation is reached. This suggests the hypothesis that after the essential interaction involving the Ia antigen, the T cell, macrophage, and the B lymphocyte, triggering terminal differentiation, the B cell may cease to synthesize the Ia polypeptides. The remaining antigen will then be gradually lost from the membrane.

A population of Ia$^+$ Ig$^-$ cells that morphologically resemble monocytes and macrophages was found in all lymphoid organs.[11] Ia$^+$ Ig$^-$ cells represented only 1 to 3% of

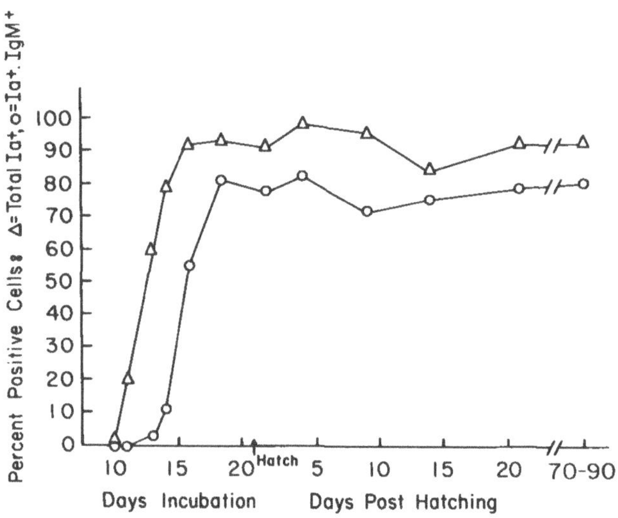

FIGURE 4. Ontogeny of total Ia^+ cells (Δ) and Ia^+ IgM^+ cells (O) in the bursa of Fabricius. Cells were stained in suspension with fluorochrome-labeled antibody to chicken μchain and alloanti-Ia antisera followed by antibody to chicken γ chain conjugated with a contrasting fluorochrome.

the mononuclear blood cells and of thymus cells, about 10% of bursal cells, 25% of bone marrow cells, and 35% of the spleen cells in mature chickens. These percentages were not diminished in juvenile agammaglobulinemic birds having no demonstrable B cells. While a portion of the Ia^+ Ig^- cell population could be identified as macrophages by their ingestion of latex particles, the remaining Ia^+ Ig^- cells in lymphoid tissues are of unknown origin and function. Also, only about 50% of the macrophages expressed detectable Ia antigens.

VI. ONTOGENY OF CELLULAR EXPRESSION

An extensive analysis of the occurrence of Ia^+ cells in various lymphoid and hemopoietic tissues during embryonic development was made using indirect immunofluorescence with alloanti-Ia antibody.[11] The appearance of Ia^+ Ig^- cells preceded or coincided with the appearance of Ia^+ Ig^+ cells in each tissue. In the bursa (Figure 4), Ia^+ cells were first detectable at a very low frequency on the 10th day of incubation. By day 13, when the first IgM^+ cells were detected at a frequency of 1.5%, Ia^+ Ig^- cells comprised 60% of the bursal population. Both Ia^+ Ig^- and Ia^+ Ig^+ populations increased in parallel until about day 18, when 90% of the bursa cells were Ia^+, and 80% also bore Ig determinants. These levels remained virtually unchanged until bursal involution occurred at about 14 weeks after hatching. When the above analysis was repeated by indirect staining with monoclonal anti-Ia antibody, the same general pattern of ontogeny was obtained, except that the percentage of Ia^+ Ig^- cells never exceeded that of Ia^+ Ig^+ cells by more than 20%. A possible reason for this difference is the lower background obtained with the monoclonal antibody, permitting easier distinction between weakly positive and negative cells.

The appearance of Ia^+ Ig^- cells in peripheral lymphoid tissues (see Figure 5) generally precedes that of Ig^+ cells by several days. Preliminary observations indicate that Ia^+ Ig^- cells, larger than thymocytes, may be present in the embryonic thymus as early as 10 days of incubation.

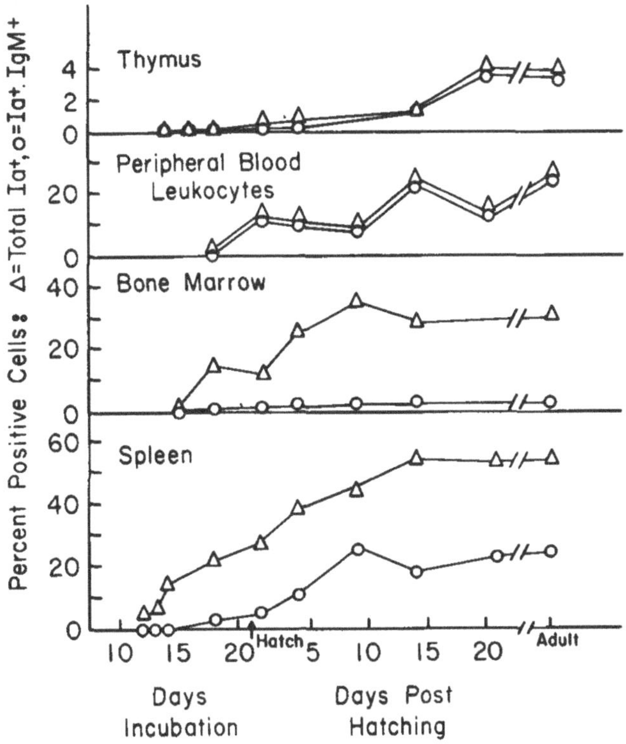

FIGURE 5. Ontogeny of total Ia⁺ cells (Δ) and Ia⁺ IgM⁺ cells (O) in various lymphoid tissues of the chicken. Staining was as in Figure 3.

The sequential appearance of Ia^+ Ig^- cells followed by Ia^+ Ig^+ cells in the bursa suggests that the latter are the progeny of the former. It is also possible that these Ia^+ Ig^- cells in the bursa and other tissues are either fully differentiated cells of the monocyte-macrophage series or stem cells from which monocytes and/or B cells are derived. However, the selective cellular driving potential of the bursal environment for differentiation and proliferation of B cells is indicated by the differential effect which removal of the organ has on the Ia^+ Ig^- and Ia^+ Ig^+ populations. Surgical bursectomy after hatching coupled with *in ovo* injection of anti-μ antibody or destruction of the bursal primordium at 3 days of incubation drastically reduces B cell levels but permits normal development of Ia^+ Ig^- cells in peripheral blood and spleen.[43] Attempts to eliminate Ia^+ cells as they arise during ontogeny by the *in ovo* injection of anti-Ia antibody were unsuccessful.[44] The reason for the failure of anti-Ia antibody to affect Ia^+ cell levels is not apparent, as the alloantibody was capable of complement-mediated cell killing. By comparison, Fairchild[45] found that fetal mice developing in the presence of high levels of maternal anti-Ia antibody to paternal Ia antigens also developed normal levels of Ia^+ cells. It is possible that sufficient levels of antibody were not obtained in either case to eliminate the Ia^+ cell development.

VII. SUMMARY AND CONCLUSIONS

The evidence accumulated thus far indicates that Aves have genes within their major histocompatibility complex that code for antigens resembling the Ia antigens of mammals. The antigens characterized thus far are glycoproteins composed of a heavy and light chain of 32,000 and 27,000 daltons, respectively. These antigens are expressed in B lymphocytes and subpopulations of cells of the monocyte/macrophage series. During ontogeny, Ia determinants can be detected on the B cell lineage as soon as they are identifiable by expression of sIgM. However, only a fraction of the fully differentiated B cells, i.e., plasma cells, bear detectable amounts of Ia determinants. These antigens appear to be responsible, in part, for stimulation in mixed lymphocyte cultures. Although the Ia genes and immune response genes are both closely linked to the B MHC genes, identity has not been established.

Because of the importance of Ia antigens in mammalian immune reactions, it is not surprising that other vertebrates with similar immune capabilities have comparable alloantigenic systems. Thus, the demonstration of a particular Ia analogue in the chicken may be suggestive of a more complex system as has been described in the mouse. One significance of Ia antigens in chickens relates to the experimental advantages the chicken offers for the study of B lymphocyte development. In this respect, the Ia antigens should provide a useful marker for analysis of the differentiation of lymphomyeloid cells. Also, reagents to alleles of Ia antigens associated with particular immune response genes should provide a practical means of selecting for disease-resistant lines of chickens.

ACKNOWLEDGMENTS

This work was supported by research grants CA 10815 and CA 16673 of the National Cancer Institute, and a grant from the W. W. Smith Foundation.

REFERENCES

1. **Pazderka, F., Longenecker, B. M., Law, G. R. J., and Ruth, R. F.,** The major histocompatibility complex of the chicken, *Immunogenetics,* 2, 101, 1975.
2. **Longenecker, M. and Mosmann, T. R.,** Structure and properties of the major histocompatibility complex of the chicken: speculations of the advantages and evolution of polymorphism, *Immunogenetics,* in press.
3. **Pink, J. R. L., Droege, W., Hala, K., Miggiano, V. C., and Ziegler, A.,** A three-locus model for the chicken major histocompatibility complex, *Immunogenetics,* 5, 203, 1977.
4. **Ziegler, A. and Pink, R.,** Structural similarity of major histocompatibility antigens on leukocytes and erythrocytes, *Immunochemistry,* 15, 515, 1978.
5. **Hala, K., Vilhelmova, M., and Hartmanova, J.,** The structure of the major histocompatibility complex of the chicken, *Adv. Exp. Med. Biol.,* 88, 227, 1977.
6. **Ziegler, A. and Pink, R.,** Chemical properties of two antigens controlled by the major histocompatibility complex of the chicken, *J. Biol. Chem.,* 251, 5391, 1976.
7. **Pink, J. R. L., Miggiano, V. C., and Ziegler, A.,** Antigens of the chicken major histocompatibility complex, *Folia Biol.,* 23, 404, 1979.
8. **Hala, K., Hasek, M., Hlozanek, I., Hort, J., Knizetova, H., and Mervartova, H.,** Syngeneic lines of chickens. II. Inbreeding and selection within the M,W, and I lines and crosses between the C, M, and W lines, *Folia Biol.,* 12, 407, 1966.
9. **Hasek, M., Knizetova, F., and Mervartova, H.,** Syngeneic lines of chickens. I. Inbreeding and selection by means of skin grafts and tests for erythrocyte antigens in Cline chickens, *Folia biol.,* 12, 335, 1966.
10. **Pink, J. R. L.,** Lectin binding products of chicken major histocompatibility complex, *Folia Biol.,* 25, 333, 1979.
11. **Ewert, D. L. and Cooper, M. D.,** Ia-like alloantigens in the chicken: serologic characterization and ontogeny of cellular expression, *Immunogenetics,* 7, 521, 1978.
12. **Ewert, D. L., Gilmour, D. G., Briles, W. E., and Cooper, M. D.,** Genetics of Ia-like alloantigens in chickens and linkage with B major histocompatibility complex, *Immunogenetics,* 10, 169, 1980.

13. **Ewert, D. L., Munchus, S., Chen, C., and Cooper, M. D.**, unpublished data, 1980.
14. **Lonai, P. and McDevitt, H. O.**, the expression of I-region gene products on lymphocytes, *Immunogenetics*, 4, 33, 1977.
15. **Cresswell, P. and Geier, S. S.**, Antisera to human B-lymphocyte membrane glycoproteins block stimulation in mixed lymphocyte culture, *Nature (London)*, 257, 1975.
16. **VanRood, J. J., van Leeuwen, A., Termijtelen, A., and Keuning, J. J.**, B-cell antibodies, Ia-like determinants, and their relation to MLC determinants in man, *Transplant. Rev.*, 30, 122, 1976.
17. **Miggiano, V. C., Birgen, J., and Pink, J. R. L.**, The mixed leukocyte reaction in chickens. Evidence for control by the major histocompatibility complex, *Eur. J. Immunol.*, 4, 397, 1974.
18. **Jones, R. F.**, Mixed lymphocyte culture and T cell mitogen responses of normal and obese strain chickens, Masters thesis, Wayne State University, Detroit, Michigan, 1977.
19. **Simonsen, M.**, MHC in a birds' eye view, 7th Int. Convocation Immunol., in press.
20. **Toivanen, A. and Toivanen, P.**, Histocompatibility requirements for cellular cooperation in the chicken: generation of germinal centers, *J. Immunol.*, 118, 431, 1977.
21. **Schwartz, R. H., David, C. S., Sachs, D. H., and Paul, W. E.**, T-lymphocyte-enriched murine peritoneal exudate cells. III. Inhibition of antigen-induced T-lymphocyte proliferation with anti-Ia antisera, *J. Immunol.*, 117, 531, 1976.
22. **Thomas, D. W. and Shewach, E. M.**, Nature of the antigenic complex recognized by T lymphocytes. Specific sensitization by antigens associated with allogeneic macrophages, *Proc. Natl. Acad. Sci. U.S.A.*, 74, 2104, 1977.
23. **Pevzner, I., Kaeberle, M. L., and Nordskog, A. W.**, Differences in antibody response between B locus blood group genotypes in Leghorns, *Poult. Sci.*, 52, 2073, 1973.
24. **Pevzner, I., Kaeberle, M. L., and Nordskog, A. W.**, Evidence that the B blood group locus in Leghorns contains an immune response region, *Genetics*, 77, 251, 1974.
25. **Pevzner, I., Nordskog, A. W., and Kaeberle, M. L.**, Immune response and the B blood group locus in chickens, *Genetics*, 80, 753, 1975.
26. **Balcarova, J., Derka, J., Hala, K., and Hraba, T.**, Genetic control of immune response to the dinitrophenol group in inbred lines of chickens, *Folio Biol.* 20, 346, 1974.
27. **Gunther, E., Balcarova, J., Hala, I., Rude, E., and Hraba, T.**, Evidence for an association between immune responsiveness of chicken to (T,G)-A--L and the major histocompatibility system, *Eur. J. Immunol.*, 4, 548, 1974.
28. **Balcarova, J., Gunther, E., Hala, K., Rude, E., and Hraba, T.**, Further evidence for the genetic control of immune responsiveness to (T,G)-A--L by the B system in chickens, *Folio Biol.*, 21, 406, 1975.
29. **Benedict, A. A., Pollard, L. W. Morrow, P. R., Abplanaip, H. A., Maurer, P. H., and Briles, W. E.**, Genetic control of immune responses in chickens. I Responses to a terpolymer of Poly (Glu^{60} Ala^{30} Tyr^{10}) associated with the major histocompatibility complex, *Immunogenetics*, 2, 313, 1975.
30. **Karakoz, I., Krejci, J., Hala, K., Blaszchizk, B., Hraba, T., and Pekarek, J.**, Genetic determination of tuberculin hypersensitivity in chicken inbred lines, *Eur. J. Immunol.*, 4, 345, 1974.
31. **Pevzner, I. Y., Trowbridge, C. L., and Nordskog, A. W.**, Recombination between genes coding for immune response and the serologically determined antigens in the chicken B system, *Immunogenetics*, 7, 25, 2978.
32. **Silver, J., Walker, L. E., Reisfeld, R. A., Pellegrino, M. A., and Ferrone, S.**, Structural studies of murine I-E and human DR antigens, *Molec. Immunol.*, 16, 37, 1979.
33. **Walker, L. E., Ferrone, S., Pellegrino, M. A., and Reisfeld, R. A.**, Structural polymorphism of β chain of human HLA-DR antigens, *Molec. Immunol.*, 17, 1443, 1980.
34. **Jones, P. P., Murphy, D. B., Hewgill, D., and McDevitt, H. D.**, Detection of a common polypeptide chain of I-A and I-E subregion immunoprecipitates, *Immunochemistry*, 16, 51, 1978.
35. **Cook, R. G., Uhr, J. W., Capra, J. D., and Vitetta, E. S.**, Structural studies on the murine Ia alloantigens II. Molecular weight characterization of the products of the I-A and I-E/C subregions, *J. Immunol.*, 121, 2205, 1978.
36. **Allison, J. P., Walker, L. E., Russell, W. A., Pellegrino, M. A., Ferrone, S., Reisfeld, R. A., Frelinger, J. A., and Silver, J.**, Murine Ia and human DR antigens: homology of amino-terminal sequences, *Proc. Natl. Acad. Sci. U.S.A.* 75, 3953, 1978.
37. **Pink, J. R. L.**, personal communication, 1980.
38. **Brogen, C.-H., Bisati, S., and Simonsen, M.**, The major histocompatibility complex antigens as lectin receptors on lymphocytes. A study by O'Farrell's two-dimensional gel electrophoresis, in *Protides of Biological Fluids, 27th Colloquium*, Peters, H., Ed., Pergamon Press, Oxford, 1979, 467.
39. **Kvist, S., Klarskog, L., and Peterson, P. A.**, Identification of H-2 and Ia-antigen analogues in several species by immunological crossreactions of xenoantisera, *Scand. J. Immunol.*, 7, 447, 1978.

40. **Okazak, W., Witter, R. L., Romero, C., Nazerian, K., Sharma, J. M., Fadly, A., and Ewert, D.,** Induction of lymphoid leukosis transplantable tumors and the establishment of lymphoblastoid cell lines, *Avian Pathol.*, 9, 311, 1980.
41. **Nowak, J. S., Vainio, O., Lassiala, O., Toivanen, P., and Gilmour, D. G.,** A rosette assay for the identification of Ia-like alloantigens on chicken lymphoid cells, *J. Immunol. Meth.*, in press.
42. **Ivani, J.,** personal communication, 1980.
43. **Ewert, D. L. and Cooper, M. D.,** Effects of embryonic bursectomy on development of Ig^+ Ia^+ cells in chickens, *Fed. Proc. Fed. Am. Soc. Exp. Biol.*, 38, 1367, 1979.
44. **Ewert, D. L.,** unpublished data, 1980.
45. **Fairchild, S. S. Cooper, M. D., and Klein, J.** Role of Ia determinants in B lymphocyte maturation, *Fed. Proc. Fed. Am. Soc. Exp. Biol.*, 33, 2152, 1979.

Chapter 2

THE Ia ANTIGENS OF THE GUINEA PIG

Reinhard Burger and Ethan M. Shevach

TABLE OF CONTENTS

ABBREVIATIONS

GL = a copolymer of L-glutamic acid and L-lysine
GT = a copolymer of L-glutamic acid and L-tyrosine
Ir = immune response
LNL = column-purified T cells from lymph nodes
PEL = column-purified T cells from peritoneal exudate cells
OVA = ovalbumin
PHA = phytohemagglutinin
DNP-GPA = 2, 4-dinitrophenyl-guinea pig albumin
PPD = purified protein derivative of tuberculin
FACS = fluorescence activated cell sorter

I. INTRODUCTION

The first immune response (Ir) gene was discovered in the guinea pig in 1963.[1] Ir genes control the ability of an animal to mount an immune response to T cell dependent antigens.[2] Linkage of Ir genes to the major histocompatibility complex (MHC) was subsequently discovered in each species studied.[3,4] The I region of the MHC was intensively analyzed in a number of laboratories in order to elucidate the function of I region products in the regulation of the immune response.

Lymphocytes were shown to express cell surface glycoproteins bearing antigenic determinants also encoded by I region genes. These molecules were subsequently termed I region associated (Ia-) antigens. In recent years, several detailed reviews on serological or chemical aspects of guinea pig Ia antigens or on the functional association of Ia antigens and Ir genes have been published.[5–7] Therefore, this article will only summarize the previously described data on the I region of the guinea pig and it will focus on a description of recent experiments applying monoclonal antibodies to guinea pig Ia antigens. This new type of reagent has allowed further serological and functional analysis of Ia antigens thereby extending previous knowledge obtained with alloantisera.

II. SEROLOGY AND CHEMISTRY OF GUINEA PIG Ia ANTIGENS

A. Ia Specificities Defined by Alloantisera and Their Strain Distribution

The serological and genetic characterization of the guinea pig MHC (the GPLA complex) has been hampered by the lack of inbred strains and of informative recombinants. Nevertheless, a combination of immunochemical analysis and functional studies has allowed an analysis of the organization of the guinea pig I region. Studies on the organization of the guinea pig MHC have primarily used 2 inbred strains, strain 2 and 13. Both strains originated from the same colony[8] and their MHC differs only in the I region. The two strains share identical classic histocompatibility antigens. The guinea pig MHC antigens were defined by utilizing alloantisera obtained either by cross-immunization with lymphoid cells of randombred guinea pigs,[9] or by cross-immunization of the inbred strains 2 and 13.[10,11] In general, the alloantisera were prepared by immunization with a homogenate of lymph node and spleen cells, approximately 100 × 10^6 cells per recipient, emulsified in complete Freund's adjuvant containing 0.5 mg/ml Mycobacterium tuberculosis H37Ra. Repeated booster injections (s.c. or i.p.) with the same cell number were given. The antisera were tested in a ^{51}Cr release assay for cytotoxicity against the appropriate target cells using normal guinea pig serum as a source of complement.[12]

One group of antigens originally described in 1972 was present in four different antigenic specificities which behaved as products of allelic genes at a single locus.[9,13] These 4 antigens, the products of the GPLA-B locus, were termed B.1, B.2, B.3, and B.4. In addition, strain 2, strain 13, BIOB, and individual outbred animals bear the antigen, S.1, which is the product of the GPLA-S locus, for which no additional alleles are known so far.

The B and S antigens resemble, in their tissue distribution and molecular characteristics, the products of murine D or K region genes and thus are the guinea pig equivalents of the H-2D/K antigens or the human HLA-A/B/C antigens. The B and S antigens are glycoproteins with a molecular weight of 40,000 daltons and are associated noncovalently with 12,000 dalton chain which has been shown to be 2-microglobulin.[14,15] Amino acid sequence data revealed significant homology of the GPLA-B and S antigens with histocompatibility antigens of mouse, rat, and man.[16]

A second group of antigens proved to be the guinea pig analogues of the murine Ia or human Ia-like antigens. Cross-immunization of strain 2 and strain 13 produced alloantisera (strain 2 anti-strain 13 and strain 13 anti-strain 2) which detected antigens with similar tissue distribution as the Ia antigens of the mouse. Strain 2 and strain 13 bear serologically identical B.1 and S.1 antigens and lack B.2, B.3, and B.4. A detailed serological and immunochemical analysis revealed that the alloantisera recognized several distinct alloantigenic specificities.[17] Two anti-13 serum represents a mixture of antibodies directed to the 3 distinct Ia specifities, Ia.1, Ia.3, and Ia.7, while 13 anti-2 serum contained antibodies to Ia.2 and Ia.4.[18] Subsequently, additional alloantisera were raised through cross-immunization of random-bred guinea pigs of known phenotype. Donor and recipients were matched for B and S antigens. The resulting antisera allowed the definition of the additional Ia-specificities Ia.5, Ia.6, and Ia.8. Antibodies to these Ia determinants are not present in 2 anti-13 or 13 anti-2 sera.[17]

Table 1 summarizes the B, S, and Ia antigens of different inbred strains of guinea pigs including some partially inbred families. Additional details of these guinea pig strains including holder, genetic traits, degree of inbreeding, etc., are listed in a recent manual.[19] Table 1 illustrates that most strains possess either the B.1 or B.3 antigen; only 2 partially inbred strains bear B.2. There is no strain available homozygous for B.4. It is also apparent from Table 1 that all strains tested so far except strain 2 share

Table 1
Strain distribution of B-, S-, and Ia-antigens

Strain	B-/S-Regions	Ia determinants
2	B.1, S.1	Ia.2,4,5,6
13	B.1, S.1	Ia.1,3,5,6,7
BIOAC	B.2 –	Ia.1,3,5,6,7
BIOAD	B.3 –	Ia.1,3,5,6,7
BIOB	B.1, S.1	Ia.1,6
BIOC	B.2 –	Ia.1,6,7,8
B/LAC	B.3 –	Ia.1,6,7,8
C4D	B.1 –	Ia.1,3,5,6,7
C2D	B.3 –	Ia,1,3,5,6,7
OM3	B.3 –	Ia.1,3,5,6,7
R9	B.3 –	Ia.1,3,5,6,7

Note: Data taken from references and unpublished results. Table 1 summarizes the B-, S- and Ia-antigens of different inbred strains of guinea pigs including some partially inbred families. Additional details of these guinea pig strains including holder, genetic traits, degree of inbreeding etc. are listed in a recent manual.[19]

I region specificities with strain 13 animals. In contrast, none of the strains possesses the Ia antigens Ia.2 or Ia.4 of strain 2.

The MHC haplotype of some strains carrying the Ia-antigens of strain 13 varies in the B locus from strain 13 and additionally in the loci for factor B, C4, and C2 of the complement system.[20] Thus, most strains possess the Ia antigens of strain 13 in combination with the B.3 or B.1 antigens. Ia.3 was always found associated with Ia.5; similarly, Ia.1 was always present together with Ia.6. A new Ia specificity was recently provisionally termed Ia.8.[17] This specificity was present in the B/LAC strain which shares the Ia.1,6 and the Ia.7 specificity with strain 13. Ia.8 resembles Ia.3,5 in its chemical structure and might represent an allele of Ia.3,5.

It is apparent from these data that in the guinea pig the serological heterogeneity in the MHC is more restricted than in other species, in particular the mouse. A possible explanation for this restricted genetic variability is that the domesticated guinea pig originates from a limited population. Guinea pigs were brought by the Spanish to Europe about 400 years ago. In wild guinea pigs caught at their natural habitat in Peru, some of the known Ia antigens were present; B or S antigens were only identifiable in about two thirds of the animals.[13]

B. Immunochemical Methods for Analysis of Guinea Pig Ia Antigens

Despite the fact that Ia antigens are readily detectable by serological methods, there are substantial problems in isolating and characterizing these cell surface glycoproteins. This is due to the fact that Ia antigens are only present in the cell membrane in trace quantities. In addition, Ia molecules are not soluble in aqueous solutions and their purification requires the use of detergents for solubilization of the membrane. Sensitive specialized methods originally developed for the analysis of mouse histocompatibility antigens were modified for the guinea pig systems.[21-24] These microanalytical techniques depend on the use of radiolabeled material. In the following, a short overview of the published methods[6,7,18] used for isolation and for structural analysis of guinea pig Ia antigens will be given.

1. Radiolabeling of Ia antigens

Two techniques are used for radiolabeling. The first represents biosynthetic or internal labeling of Ia antigens using ^{3}H, ^{14}C, or ^{35}S labeled amino acids or sugars. Due to a high synthesis rate of Ia antigens in short-term cultures, efficient incorporation of the radioactive material into the membrane-associated Ia molecules is obtained. Generally, lymphocytes from lymph nodes are used as a source of Ia antigens. The alternative labeling procedure involves radioiodination of cell surface associated Ia molecules[25] by lactoperoxidase catalyzed iodination. Results obtained with this method differ from those obtained with the first labeling technique as discussed below, probably due to problems of accessibility or orientation of the Ia molecule in the membrane.

2. Solubilization of Ia Antigens

The labeled Ia antigens in the cell membrane are solubilized by the nonionic detergent Non-idet 40 (NP 40). This method preserves the antigenicity of the Ia antigens. The supernatant after ultracentrifugation is used for further purification.

3. Purification by Affinity Chromatography

A partial purification is achieved by chromatography on columns of immobilized lentil (*Lens culinaris*) lectin. The carbohydrate-containing Ia glycoproteins are bound through their glucose and mannose residues to the lectin. The nonbound material is removed by washing and the bound material is eluted with α-methyl mannoside. The majority (70 to 80%) of the applied Ia antigenic material can be recovered in a significantly enriched and purified form. For certain purposes, further purification might be required, because at this step, there are still contaminating glycoproteins present. Highly purified Ia antigens can be obtained in relatively large amounts by an additional step of immunoabsorbent purification using immobilized antibodies to Ia antigens for affinity chromatography.[26]

4. Immunoprecipitation

Anti-Ia antibodies are added to the radiolabeled Ia antigen preparation. The immune complexes which are formed do not precipitate and are isolated from the fluid phase by precipitation with heterologous anti-immunoglobulin sera or, alternatively, by addition of fixed, protein A bearing *Staphylococcus aureus* Cowan I strain. The second technique has several advantages over the double antibody precipitation.[21,27]

5. Electrophoresis in Sodium-Dodecyl-Sulfate-Polyacrylamide Gels (SDS-PAGE)

A modification of the discontinuous SDS-PAGE system is used for determination of the subunit composition and of the molecular weight of the precipitated Ia antigens.[14,22]

C. Structural Characteristics of Guinea Pig Ia Antigens

1. Molecular Weight and Association of Ia Subunits

The molecular characteristics of the guinea pig Ia antigens were determined by immunoprecipitation of radiolabeled Ia molecules with alloantisera followed by SDS-PAGE analysis under reducing and nonreducing conditions. A representative electropherogram is shown in Figure 1.

The molecules bearing the Ia determinants could be divided into three distinct groups which differ in their subunit structure.[7,18] The first group consists of molecules composed of 2 components, a larger 33,000 dalton chain (α-chain) and a smaller, 25,000 dalton chain (β-chain) in *noncovalent* association. The Ia.2 bearing molecule of strain 2, the Ia.3,5 bearing molecule of strain 13, and the Ia.8 bearing molecule of the B/

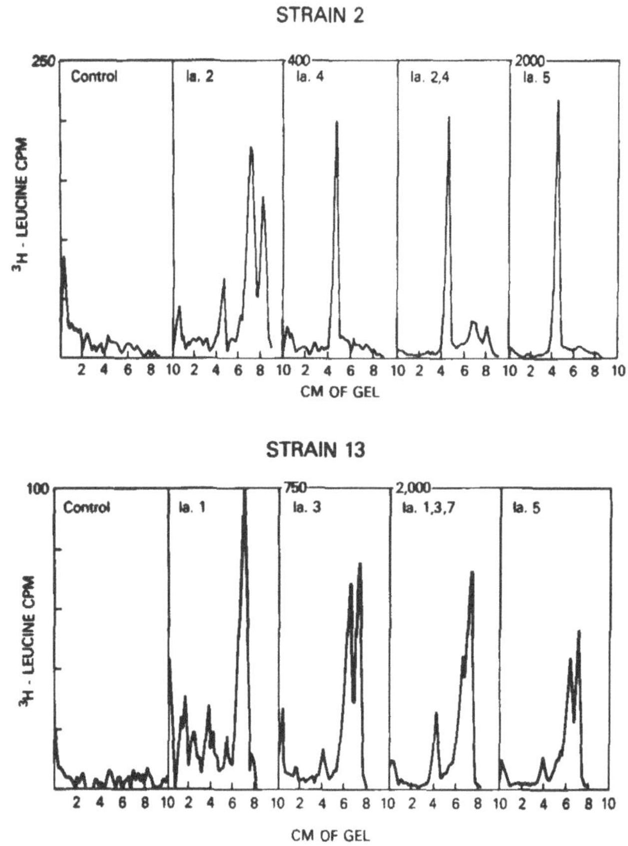

Figure 1. SDS-PAGE of guinea pig Ia antigens. Typical electropherograms obtained in SDS-PAGE (10% gels, 10 cm, under nonreducing conditions) of immunoprecipitated, biosynthetically labeled Ia antigens prepared from strain 2 and strain 13 lymph node cells. (A) Strain 2 lymph node cells. (B) Strain 13 lymph node cells.

LAC strain have these characteristics. A second group of Ia antigens are borne on molecules composed of 2 chains with molecular weights of 33,000 and 25,000 daltons which are linked *covalently* by disulfide bonds. The Ia.4,5 bearing molecule of strain 2 and the Ia.7 bearing molecule of strain 13 fall into this category. A third group consists of the Ia.1,6 bearing molecule of strain 13 which appears to be a single chain of 26,000 to 27,000 daltons. However, we cannot exclude the possibility that the Ia.1,6 determinants are borne on two noncovalently linked chains of very similar molecular weight. The covalent linkage of some of the guinea pig Ia subunits is unusual in that noncovalent association of the α and β chains of the Ia molecule has been observed in most other species studied.

Molecules bearing specific Ia determinants were identified either with antisera specific for a given determinant, or if no monospecific antisera were available (e.g., anti-Ia.7), through a combination of serological and physicochemical methods. Sequential immunoprecipitation followed by the SDS-PAGE proved to be especially useful.[22] In this technique, radiolabeled Ia molecules bearing one antigenic determinant are first completely removed by immunoprecipitation. The remaining supernatant is then tested for the presence or absence of molecules bearing a second antigenic determinant. Sequential precipitation allows one to distinguish whether a single molecular species bears several Ia antigenic determinants, or whether they are located on different molecules. This technique allowed us to demonstrate that Ia.3 and Ia.5 are borne on a single molecule, which in turn can be distinguished from the Ia molecule bearing both the Ia.1 and Ia.6 determinant and from a third molecular species bearing the Ia.7 specificity. The molecular characteristics of a molecule bearing a given Ia specificity were identical in most strains bearing the determinant. For example, Ia.7 was always found on a covalently but never on a noncovalently linked 58,000 dalton molecule in all inbred and outbred animals studied. One exception was the Ia.5 specificity which is shared between strain 2 and strain 13 and is found on structurally different molecular species. In strain 2, Ia.5 was present on the same 58,000 dalton molecule as Ia.4 composed of 2 covalently linked chains, whereas in strain 13 it was present on the same molecule as Ia.3 composed of 2 noncovalently linked chains.[18] In strain 2, Ia.2 is borne on a molecule which can be distinguished from the Ia.4 and Ia.5 bearing molecule. The molecular species bearing the Ia.6 determinant has not yet been clearly identified in strain 2.

The differences in molecular weight and in the association of the 2 chains of an Ia molecule allowed one to identify a single Ia specificity even if a given antiserum was directed to more than a single specificity. For example, we were unable to raise an antiserum monospecific for Ia.7, but an anti-Ia.1.7 was raised. Because Ia.7 is found on a molecule with a molecular weight of 58,000 daltons under nonreducing condition, SDS-PAGE after immunoprecipitation with anti-Ia.1,7 readily distinquished Ia.7 from Ia.1, which has a molecular weight of 26,000 daltons under both reducing and nonreducing conditions.

The two-chain structure of guinea pig Ia antigens was recognized using internally labeled Ia antigens. After surface iodination of the Ia.3,5 or Ia.2 bearing molecules, only the 25,000 dalton β chain, but not the 33,000 dalton component, showed efficient labeling.[28] Apparently, the larger α chain of membrane-associated Ia antigens is poorly radioiodinated, probably due to inaccessibility of the α chain to the lactoperoxidase. The selective labeling is not due to the absence of tyrosine residues required for lactoperoxidase catalyzed radioiodination, because tyrosine was demonstrated in both chains by biosynthetic labeling with ^{3}H-tyrosine.[29]

So far, no association of individual Ia determinants with a single chain of the two-chain molecule has been described. In addition, in the guinea pig, no evidence has as yet been obtained for the presence of the so-called invariant chain which is associated in the murine system with I-A and I-E antigens from several different haplotypes.[30]

2. Carbohydrate

The carbohydrate content of guinea pig Ia antigens has not been intensively studied. In the guinea pig system, there is no indication that carbohydrates contribute significantly to the Ia antigenic specificity. Biosynthetic studies with ^{3}H-fucose show labeling of all Ia molecules. Carbohydrate composition influences the general properties of individual Ia molecules. In strain 2, the Ia.2 bearing molecule binds very weakly to lentil lectin whereas the Ia.4,5 molecule is bound efficiently.[14]

3. Isoelectric Focusing and Sequence Data

Isoelectric focusing of radiolabeled strain 2 Ia antigens after immunoprecipitation gave a discrete banding pattern without any indication of microheterogeneity as it is seen with immunoglobulin.[6] With microsequencing techniques, partial N terminal amino acid sequences of the β chains were obtained from several Ia antigens.[31] There was a moderate homology between the β chains of Ia.4,5 of strain 2 and Ia.7 of strain 13, both representing Ia molecules with covalently linked α and β chains. Less homology was present between the β chains of Ia.3,5 and Ia.7 of strain 13, reflecting the differences in subunit structure. Some homology to the human HLA-DR analogue was present, but virtually none to the β chains of three murine Ia antigens.

D. Serological and Immunochemical Analysis of Guinea Pig Ia Antigens with Monoclonal Antibodies

1. Rationale for the Production of Monoclonal Anti-Ia Antibodies

Alloantisera directed to Ia antigens have proven to be a powerful tool in the serological and functional analysis of Ia antigens. The use of anti-Ia alloantibodies has yielded a large body of information about the molecular nature, cellular distribution, and functional expression of Ia antigens on lymphocytes and macrophages. Nevertheless, by their very nature, the use of alloantisera is limited for several reasons. Alloantibodies to Ia antigens are obtained by cross-immunization of different strains of guinea pigs and therefore are only able to recognize determinants which are polymorphic within that species. Different alleles of Ia antigens have to be expressed at a given locus as a prerequisite for successful production of alloantibodies. Determinants present in all strains are not recognized by cross-immunization. In the guinea pig, only a limited number of genetically defined strains is available thereby restricting the genetic heterogeneity and the possible strain combinations for immunizations. In practical terms, alloantisera have several disadvantages which often cannot be avoided. The antisera are often weak, vary in titer from batch to batch, and are available only in limited quantities. In addition, in the guinea pig where no truly congenic strains are available, even apparently monospecific antisera may comprise a mixture of antibodies recognizing several MHC or non-MHC coded antigens and they also may vary in isotype, biological properties, and in affinity.

Several groups have reported the production of conventional heterologous antisera to MHC antigens and have also succeeded in obtaining alloantigen-specific xenoantisera.[32-35] Xenoimmunization as a rule provides antisera of higher potency because antigenic differences across the species elicit a broader and stronger immune response than polymorphic variants of homologous antigens within a species. Xenoimmunization has the potential to produce antisera recognizing proteins which exhibit little or no genetic heterogeneity within a given species, i.e., do not represent polymorphic alloantigens. The wide range of anti-genic determinants recognized in xenoimmunization usually results in serologically complex antisera which require extensive absorptions.

The application of the hybridoma technique introduced by Kohler and Milstein[36,37] creates a tool to overcome these problems. Hybrid cell lines obtained by fusion of myeloma cells with spleen cells from mice immunized previously with antigens serve as a source of monoclonal antibodies. If a sufficiently selective screening system is available, the broad xenogeneic response can be dissected by appropriate techniques into individual clones secreting antibodies directed to a single determinant in the complex mixture used for immunization. Thus, a nonpurified antigen can be used for the production of specific antibody to an individual component. The secreted antibody is directed to a single antigenic determinant and can be obtained in large amounts.

It seemed quite feasible to use the mouse for the production of monoclonal antibodies, because guinea pig Ia antigens were previously shown to be immunogenic in other species. Antibodies to Ia antigens were present in heterologous antisera obtained after immunization of the rabbit or burro with Ia-bearing guinea pig lymphocytes.[38] We, therefore, applied the hybridoma technique for the production of monoclonal antibodies to guinea pig Ia antigens with the aim of producing reagents which allow further serological and immunochemical analysis and subsequently more detailed functional studies of guinea pig Ia antigens.

Table 2
REACTIVITY OF THE MONOCLONAL ANTI-Ia ANTIBODIES IN BINDING AND CYTOTOXICITY TESTS, IN FACS ANALYSIS AND IMMUNOPRECIPITATION EXPERIMENTS

	22C4	25E3	25E11	27E7
Binding				
EN-L2C	+	+	+	+
BZ-L2C	−	−	−	−
Str.2 spleen	+	+	+	+
Str.13 spleen	+	−	−	+
Cytotoxicity				
EN-L2C	+	+	+	+
BZ-L2C	−	−	−	−
G13	+	−	−	−
Str.2 spleen	+	+	+	+
Str.13 spleen	+	−	−	−
FACS analysis				
Str.2 PEL	+	+	+	+
Str.2 thymus	+	+	+	+
Str.13 PEL	+	−	−	+
Str.13 thymus	+	+	−	+
Immunoprecipitation				
Str.2 Ia antigens	Ia.2; Ia.4,5	?	?	Ia.2
Str.13 Ia antigens	Ia.1,6; Ia.3,5; Ia.7	None	None	Ia.1,6; Ia.3,5; Ia.7

From Burger, R., Clement, L., Schroer, J., Chiba, J., and Shevach, E. M., *J. Immunol.*, 126, 32, 1981. With permission.

We summarize below the methods for the production and the serological characterization of four independently derived monoclonal antibodies to Ia antigens. Details are given elsewhere.[39,40]

2. Production of Monoclonal Antibodies to Guinea Pig Ia Antigens

BALB/c mice were immunized with Ia-bearing EN-L2C leukemia cells of strain 2 guinea pigs suspended in complete Freund's adjuvant. After two booster injections the spleen cells were fused according to standard procedures to the nonsecretor mouse myeloma line NS-1.[41] Supernatants from growing hybrids were screened for the presence of anti-Ia antibodies in a binding assay using ^{125}I-labeled F(ab′)$_2$ anti-mouse Ig. Hybrids reacting selectively with Ia positive EN-L2C cells, but not with its Ia negative variant BZ-L2C, were cloned and ascitic fluid was produced as a source of antibody. Four of the hybrids showing this selective reactivity were analyzed in detail and were shown to identify Ia antigens. Three of these antibodies (22C4, 25E11, 27E7) were of the IgG$_1$ isotype; 25E11 was an IgM antibody.

3. Binding Assays with Various Target Cells

The four monoclonal antibodies were analyzed in binding and cytotoxicity tests for their reactivity with lymphoid cells. Table 2 shows the reactivity pattern with the various cell types. In binding assays, all monoclonal antibodies clearly reacted with the appropriate target cells up to a dilution of 1/200,000. Figure 2 shows a representative

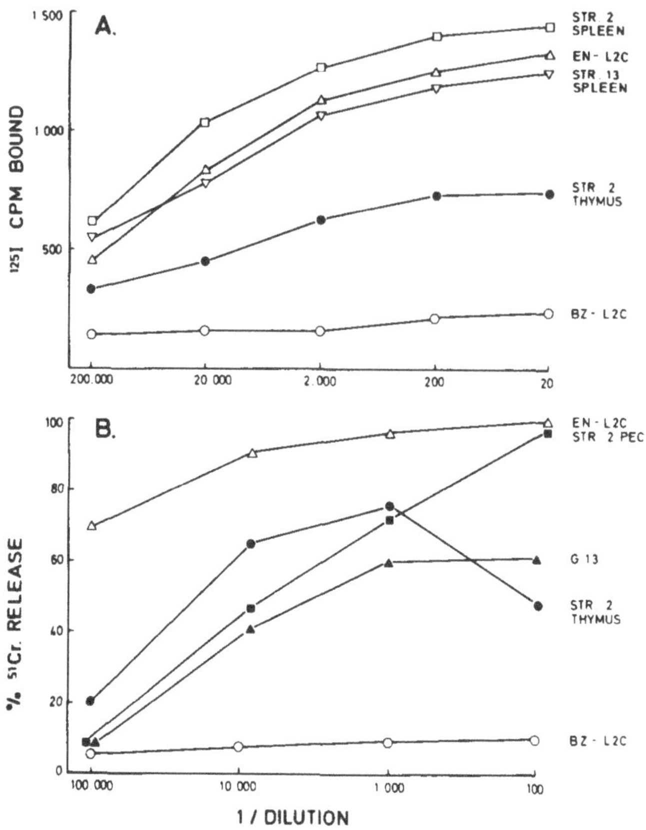

Figure 2. Reactvity of monoclonal anti-Ia antibodies with different cell types in binding (A) and in cytotoxicity assays (B). (A) 4 × 10^5 cells were incubated in a total volume of 50μℓ with ascitic fluid at the indicated dilutions and washed. After incubation with ^{125}I-labeled $F(ab')_2$ rabbit anti-mouse Ig, cell bound radioactivity was determined. Nonspecific binding obtained with an ascitic fluid containing nonrelevant antibodies was in the range of 150 to 250 cpm. EN-L2C: Ia-positive B cell leukemia of strain 2; BZ-L2C: Ia-negative variant of EN-L2C. (B) Percent ^{51}Cr-release by monoclonal antibodies at the indicated dilutions of ascitic fluid in the presence of rabbit complement. The release induced by control-ascites and complement was less than 10% with all cells tested. All antibodies failed to induce ^{51}Cr-release from the Ia-negative BZ-L2C cell line.

result obtained on strain 2 and strain 13 cells with 1 of the monoclonal antibodies, 22C4. Strongest binding was observed with EN-L2C leukemia cells and spleen cells. Antibody binding to thymocytes was also seen. Binding to the Ia-negative BZ-L2C cells did not occur above the minimal range obtained with unrelated control ascites.

EN-L2C cells, a B-cell leukemia of strain 2 origin, were used for immunization prior to the fusion. Therefore, antibodies to Ia antigens could be directed either to alloantigenic determinants expressed only on strain 2 Ia antigens, or the antibodies might recognize nonpolymorphic or "common determinants" shared between strain 2 and strain 13. A comparison of the binding to cells of the two different strains is shown in Table 2. Two of the anti-Ia antibodies, 22C4 and 27E7, reacted with cells of both strains, indicating that they identified common or framework determinants. In contrast, antibodies 25E3 and 25E11 reacted only with strain 2 cells and were therefore apparently specific for strain 2 alloantigens. None of these antibodies reacted with human cells bearing HLA-DR antigens.

4. Analysis of Specificity by Competitive Binding

In an attempt to establish identity or partial identity of the determinants recognized by the monoclonal antibodies with the determinants recognized by alloantibodies, binding inhibition experiments were performed. Strain 2 or strain 13 cells were first pretreated with 13 anti-2 (anti-Ia.2,4) or with 2 anti-13 (anti-Ia.1,3,7) alloantisera and subsequently binding of the monoclonal antibodies was determined. When strain 2 cells were preincubated with 13 anti-2 serum, a marked reduction in the binding of 25E3 and 25E11 was observed. Reactivity of 22C4 and 27E7 with strain 2 cells was unaffected by pretreatment with 13 anti-2 serum. No reduction in the binding of 22C4 and 27E7 to strain 13 spleen cells was seen following pretreatment of strain 13 spleen cells with 2 anti-13 serum. These results suggest that antibodies 25E3 and 25E11 recognize determinants which are at least partially identical to those recognized by 13 anti-2 serum. Both 22C4 and 27E7 appear to be directed to common or framework determinants present on Ia molecules of both inbred strains.

5. Cytotoxic Activity for B and T Cells and for Macrophages

Analysis in cytotoxicity assays using rabbit complement gave results similar to the binding tests. All four antibodies were cytotoxic at high titers. The reactivity pattern is shown in Table 2 and Figure 2. All antibodies lysed the Ia positive EN-L2C cells, but not the Ia negative BZ-L2C. Thymocytes, upseparated lymph node cells, purified lymph node T cells, oil induced peritoneal exudate cells, or adherence purified macrophages of strain 2 guinea pigs were killed by all four antibodies. In contrast, only 22C4 gave significant lysis of these cell types from strain 13 animals. Antibody 27E7 clearly binds in radioimmunoassays or in FACS analysis to the corresponding cells of both strains and immunoprecipitates strain 13 Ia antigens. However, this antibody was only capable of lysing cells from strain 2 but not from strain 13 animals. A possible explanation for this differential ability to 27E7 to lyse cells of the two strains is that the corresponding determinant is expressed at a higher density on strain 2 cells than on strain 13 cells. This could lead to a more efficient activation of complement because of the higher frequency of adjacent IgG molecules which are required for activation of C1. In radioactive-binding assays, however, no preferential binding to strain 2 cells was found.

Binding and cytotoxicity assays both revealed a strong reactivity of the monoclonal antibodies with strain 2 purified T cells or with thymocytes. The majority of these cells were also killed. A detailed analysis and discussion of the reactivity with thymocytes and T cells is given in Section III.B. which summarizes data obtained with the FACS. The efficient killing of these populations is in marked contrast to the moderate degree of killing obtained with alloantisera. A similar phenomenon is seen with macrophages where 60% of cells were killed by the monoclonal antibodies, but only 15 to 25% were killed by alloantisera. One reason for the difference between the results obtained with the alloantisera and with the monoclonals is the high titer of antibodies in the ascitic fluid, exceeding by far the titers obtained in conventional alloantisera.

6. Immunoprecipitation of Guinea Pig Ia Antigens with Monoclonal Antibodies

The assumption that the monoclonal antibodies were directed to Ia antigens was based mainly on the absence of reactivity with the Ia negative leukemia variant BZ-L2C. Immunoprecipitation experiments of radiolabeled cell membrane associated Ia molecules confirmed this conclusion. ^{3}H-labeled or ^{125}I-labeled Ia antigens were immunoprecipitated with the individual monoclonal antibodies followed by *S. aureus* Cowan I or by rabbit anti-mouse Ig. The electropherograms obtained with the precipitates in SDS-PAGE under reducing or nonreducing conditions were compared to the patterns obtained with alloantisera. In addition, sequential immunoprecipitation experiments were performed in order to establish whether alloantisera and the monoclonal antibodies were recognizing the same molecular species.

Antibodies 22C4 and 27E7 precipitated the Ia.3,5 bearing molecule and the Ia.7 molecule of strain 13. Pretreatment of ^{3}H-labeled Ia antigens from strain 13 with either one of the antibodies removed all molecules capable of reacting with 2 anti-13 alloantisera and vice versa. Therefore, antibodies 27E7 and 22C4 also react with the Ia.1,6 bearing molecule. Antibody 22C4 also precipitated the Ia.2 and Ia.4,5 bearing molecules of strain 2. Antibody 27E7 reacted only with Ia.2, but not Ia.4,5 of strain 2, as shown in sequential immunoprecipitation (Figure 3). Radioiodinated strain 2 Ia antigens proved to be more easily detected in immunoprecipitation experiments than internally labeled material. Antibodies 25E3 and 25E11, which did not react with strain 13 cells in serological assays, failed to precipitate strain 13 Ia antigens. However, antibodies 25E3 and 25E11 were also only marginally effective in immunoprecipitation studies with strain 2 Ia antigens despite strong reactivity in all serological or functional assays.

The reactivity patterns of the monoclonal antibodies is summarized in Table 2. It is apparent that antibodies 22C4 and 27E7 are directed to common or framework determinants present on most (27E7), if not all (22C4), Ia molecules of strain 2 and strain 13. The common determinant recognized by 27E7 is different from the one recognized by 22C4 because 27E7 does not react with the Ia.4,5 molecule which is recognized by 22C4. Although monoclonals 25E3 and 25E11 failed to precipitate strain 2 Ia antigens, it is likely that these antibodies are directed to an alloantigenic determinant of strain 2 Ia antigens. The binding of both monoclonals to strain 2 cells was inhibited by pretreatment with 13 anti-2 alloantiserum, and both exerted a specific inhibitory effect on the Ia.2 linked T cell response to the polymer GL (see Section VI).

These serological and immunochemical data underline the discriminatory capacity of the hybridoma technique. The antibodies were obtained after immunization of mice with whole guinea pig cells which represent a complex and heterogeneous immunogen. Nevertheless, the monoclonals were not only specific for histocompatibility antigens, but were even able to distinquish at least three different determinants on individual Ia molecules. Thus, the Ia.2 molecule of strain 2 bears 2 different common determinants, which previously could not have been defined by alloimmunization. At least one additional strain-specific determinant, recognized by 25E3 and probably also by 25E11, is also present on the Ia.2 molecule. This determinant is apparently identical or partially identical to the alloantigenic determinant of Ia.2 recognized by a 13 anti-2 alloantiserum. According to a recent proposal about the nomenclature of determinants recognized by monoclonal antibodies, the determinants recognized by the monoclonal antibodies 22C4, 25E3, 25E11, or 27E7 are designated Ia.m1, Ia.m2, Ia.m3, or Ia.m4, respectively. The ability to detect different antigenic sites on an individual molecule has provided the basis for an attempt to study the contribution of individual Ia-determinants on Ia/Ir gene function at the macrophage-T cell level in vitro (see Section VI).

III. CELLULAR DISTRIBUTION OF GUINEA PIG Ia ANTIGENS: ANALYSIS WITH ALLOANTISERA AND MONOCLONAL ANTIBODIES

A. Tissue Distribution

The cellular distribution of Ia antigens was studied both by serological and immunochemical methods. The availability of the monoclonal murine antibodies to guinea

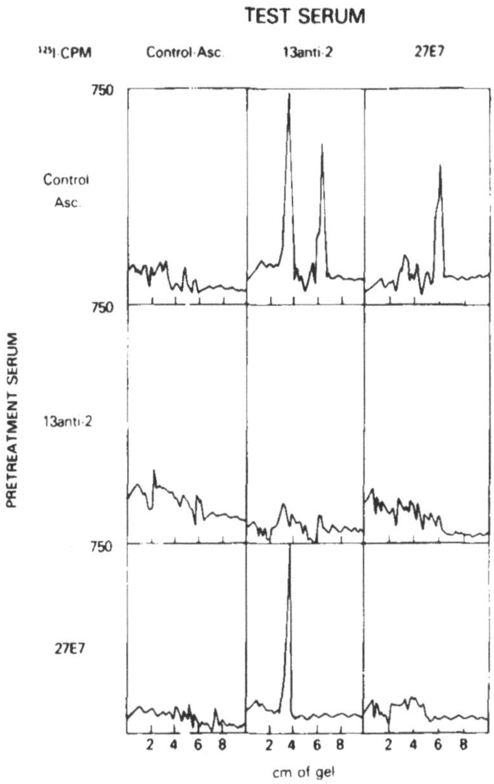

Figure 3. Sequential immunoprecipitation of ^{125}I-labeled Ia antigens of strain 2. Upper Panel: After pretreatment with control-ascites, 13 anti-2 serum precipitates the 58,000 dalton molecule bearing the Ia 4.5 determinants and the 25,000 dalton chain of the Ia.2 bearing molecule; the 33,000 dalton chain of Ia.2 is not radiolabeled by the lactoperoxidase technique. 27E7 only precipitates the 25,000 chain. Middle: Pretreatment with 13 anti-2 removes all material recognized by 27E7. Lower Panel: Pretreatment with 27E7 removes the 25,000 dalton chain recognized by 13 anti-2 but not the 58,000 dalton chain. Therefore, 27E7 recognizes the Ia.2 bearing molecule, but does not react with the Ia.4,5 bearing molecule of the strain 2 I-region. (From Burger, R., Clement, L., Schroer, J., Chiba, J., and Shevach, E. M., *J. Immunol.*, 126, 32, 1981. With permission.)

pig Ia antigens which were described in the previous section facilitated the analysis.[42] These antibodies have proven to be very useful in indirect immunofluorescence studies because the second (sandwich) antibody is directed against the mouse immunoglobulin and exhibited in our hands little, if any, cross-reactivity for guinea pig immunoglobulin. Thus, we could readily analyze B cells or B cell-containing cell preparations for the presence of Ia determinants.

Ia antigens were demonstrated with alloantisera on lymphoid cells, but were absent on a number of carcinogen or viral induced hepatomas or sarcomas.[43] Ia antigens were present in a relatively high density on B cells and on the EN-L2C B leukemia cells as demonstrated by cytotoxicity and absorption studies.[12,42,44] Cytotoxicity assays showed the presence of Ia antigens on a subpopulation of 15 to 25% of purified macrophages.[45] Indirect immunofluorescence studies with B cell and macrophage cell populations using alloantisera were difficult because of the reasons described above. Preliminary results obtained by analysis with the sensitive fluorescence activated cell sorter (FACS) in combination with monoclonal anti-Ia antibodies showed that a much higher proportion of macrophages bears Ia antigens and makes it questionable whether truly Ia-negative macrophages exist. The so-called Ia-negative macrophages, as defined in cytotoxicity assays,[45] might actually represent macrophages bearing Ia antigens at low density. Purified Kupffer cells were also shown to bear Ia antigens.[46]

Detailed analysis was performed on the expression of Ia antigens on epidermal Langerhans cells. Ia antigens were found to be selectively expressed on Langerhans cells but not on melanocytes or keratinocytes.[47,48] Studies of a transplantable granulocytic leukemia induced in strain 13 guinea pigs demonstrated the presence of Ia antigens on the leukemic myeloblasts.[49]

The reported studies on the expression of Ia antigens on the different cell types initially used serological techniques, i.e., cytotoxicity assays, indirect immunofluorescence, and absorption experiments for the demonstration of Ia antigens. It should be mentioned that most of these reports were confirmed by immunoprecipitation of biosynthetically labeled Ia antigens from cultures of purified cell populations.[50] Therefore, a putative passive absorption of external Ia molecules, e.g., from serum, can be excluded. In the mouse and man, evidence has been obtained for absorption of Ia antigens from serum[51] or for exchange of Ia antigens between different cells.[52]

Chemical characterization and comparison of Ia antigens prepared from T lymphocytes, B lymphocytes, and macrophages revealed variation in affinity for lentil lectin.[6,22,53] The majority of B cell Ia antigens were bound to the lectin, whereas T cell Ia antigens could be divided in approximately equal proportions into binding and nonbinding fractions. Ia antigens derived from macrophages did not bind to lectin. The differential behavior may be explained by variation in the carbohydrate moieties or by differential susceptibility to enzymatic action during purification. Isoelectric focusing of Ia antigens revealed discrete banding patterns for each cell type without indication for a more extensive heterogeneity in one of the cell types.[50]

B. Expression of Ia Antigens on Nonactivated T Cells and Thymocytes: FACS Analysis with Monoclonal Antibodies

Because of the critical regulatory role of Ir genes on T lymphocyte functions, it was assumed that Ir genes/Ia antigens would be functionally expressed in T lymphocytes. The initial attempts to demonstrate Ia antigens on murine T cells gave conflicting results. At present, most investigators agree that both murine and human Ia antigens are expressed on a minor subpopulation of T cells and the majority of peripheral T cells are Ia negative. However, Ia antigens in these two species can be readily demonstrated on activated T cells after stimulation with mitogens, antigens, or alloantigens.[54-56]

In contrast, in the guinea pig, Ia antigens were easily demonstrated on normal peripheral T cells utilizing alloantisera in cytotoxicity tests combined with absorption and immunochemical tests.[7,12] The monoclonal antibodies were used to analyze in quantitative terms the distributions of Ia antigens on purified T cells and thymocytes.[42] Nylon-wool purified peritoneal exudate T cells, lymph node cells, or thymocytes were used for indirect immunofluorescence with the monoclonal antibodies. Ascitic fluid served as a source of antibody; background staining was assayed by using ascitic fluid containing an unrelated antibody. Staining was performed utilizing FITC-conjugated $F(ab')_2$ anti-mouse Ig as a sandwich reagent. For cytofluorographic analysis, a FACS II was used. A representative experiment is shown in Figure 4. Antibody 27E7, directed

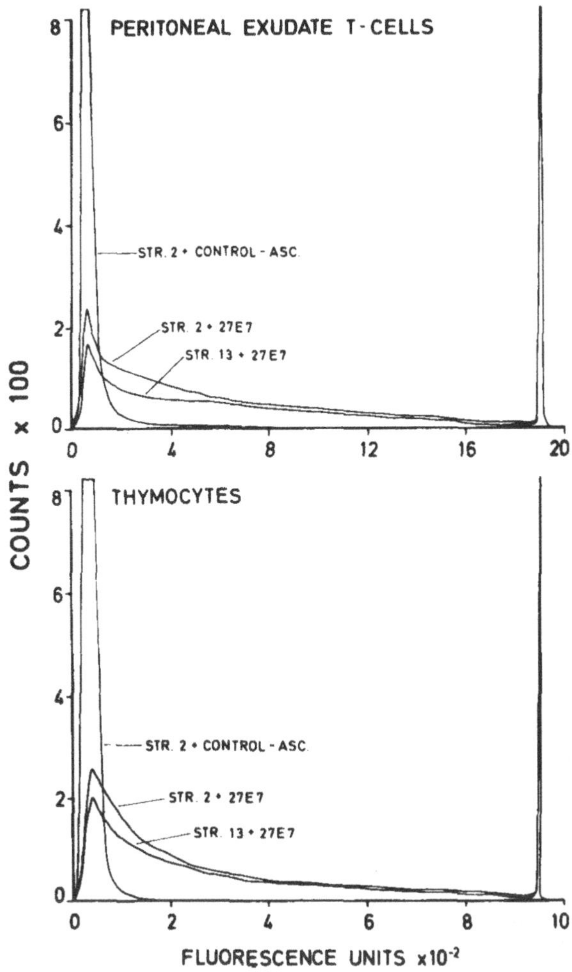

Figure 4. FACS analysis of peritoneal exudate T cells and thymocytes with monoclonal anti-Ia antibody 27E7. Strain 2 or strain 13 cells were incubated with ascitic fluid at a dilution of 1:100, washed and incubated with FITC labeled rabbit IgG anti-mouse Ig. A FACS II was used for cytofluorographic analysis. Background staining was obtained with control-ascites.

to a common determinant shared between strain 2 and strain 13 Ia antigens was used for staining. Seventy to eighty percent of the T cells and 80 to 90% of the thymocytes of both strains were Ia-positive in this sensitive assay. A similar result was obtained in both strains with antibody 22C4.

The strain 2 alloantigen-specific antibodies 25E3 and 25E11 also stained the majority of strain 2 T cells and thymocytes, but did not react with strain 13 T cells. An unexpected finding was obtained with antibody 25E3. This antibody is apparently directed to a determinant of the Ia.2 antigen of strain 2 as described above (see Section II.D.). In addition, in functional T cell proliferation assays in vitro, this antibody selectively inhibited the response of (2 × 13) F1 animals to GL (see Section VI.B.C.). Thus, it behaved similarly to a monospecific alloantiserum directed to Ia.2 which also specifically inhibits the GL response controlled by an Ia.2 linked Ir gene. Antibody 25E11 failed to react with strain 13 thymocytes as expected from its selective reaction in binding and cytotoxicity assays with strain 2 T cells and thymocytes. However, antibody

25E3 clearly reacted with strain 13 thymocytes, but failed to stain strain 13 peripheral T cells. Therefore, strain 13 thymocytes, but not mature strain 13 T cells, express a determinant which is identical to, or at least crossreacts with, an alloantigenic determinant of strain 2. The significance of this finding is not yet known. It should be emphasized that we have used a xenogeneic monoclonal antibody directed to an alloantigenic determinant. Similar phenomona might be present in other species, but would not be detected unless an appropriate xenogeneic reagent like 25E3 is used. Alloantisera to mouse Ia or even conventional monoclonal anti-Ia antibodies could not be raised to such a determinant because it would require an immune reaction against an antigen present in the thymus of the immunized animal. It might be premature to speculate on the importance of the expression of "wrong" Ia antigens on thymocytes but not peripheral T cells, but it is possible that it relates to selection processes occurring in the course of T cell differentiation in the thymus. Further analysis of the significance of this phenomenon is required.

The Ia bearing T cells and thymocytes proved to be at least partially identical with cell populations bearing T cell differentiation antigens as defined by monoclonal anti-T cell antibodies. These data demonstrate that in the guinea pig the majority of non-activated peripheral T cells and thymocytes bear Ia antigens. The quantitative analysis of the expression of T cell Ia was facilitated through the use of a combination of well-defined monoclonal antibodies and the sensitivity of the FACS. These results are in contrast to the findings in the human and the mouse where Ia antigens are predominantly expressed on activated T cells. On the whole, however, with the exception of T cells, the cellular distribution of Ia antigens in the guinea pig resembles the distribution present in other species. Cell sorter studies using alloantisera and purified T cells gave results similar to those obtained with monoclonal antibodies. The majority of peritoneal exudate or lymph node T cells proved to be Ia positive.[42] Depending on the strain studied, 40 to 80% of thymocytes were Ia positive. Absorption of 13 anti-2 alloantiserum with Ia bearing cells of the strain 2 B cell leukemia, EN-L2C, but not with the Ia-negative variant, BZ-L2C, removed all reactivity for T cells, supporting the assumption of identity of Ia antigens on B and T cells.

IV. GENETIC ORGANIZATION OF THE GUINEA PIG I REGION: Ir GENES AND Ia-Ir ASSOCIATION

In the original description of Ir genes by Benacerraf and co-workers,[1] it was noted that individual Hartley guinea pigs had the ability to mount an immune response to the synthetic amino acid polymer 2,4-dinitrophenyl-poly-L-lysine (DNP-PLL). The presence or absence of an immune response could be observed in assays measuring humoral or cell-mediated immunity. A dominant, autosomal gene, (the PLL-gene), was found by breeding studies to control the ability to respond to PLL and was also present in strain 2 but not in strain 13 guinea pigs. In the mouse, linkage of similar Ir genes to the H-2 complex was found and subsequently linkage of Ir genes to the guinea pig MHC was demonstrated by analysis of backcross offspring.

A number of other Ir genes have been demonstrated in the inbred strains 2 and 13.[5,19,57] Most antigens, the response to which is controlled by Ir genes, represent synthetic polymers of simple chemical structure and of restricted antigenic heterogeneity. However, MHC linked Ir genes were also found to control complex immune phenomena like the susceptibility to autoimmune thyroiditis in strain 2 guinea pigs or to autoimmune tubulointerstitial nephritis, which can be induced in strain 13 but not in strain 2 animals.[58,59]

The analysis of the association of Ia antigens and Ir gene function was the subject of a recent, detailed review[5] and, therefore, will be only briefly summarized. Alloantisera obtained by cross-immunization of strain 2 and strain 13 guinea pigs had a selective inhibitory effect on the proliferative response of (2 × 13) F1 T cells.[60] The antisera specifically inhibited the T cell response to antigens controlled by Ir genes which were linked to the Ia antigens against which the alloantisera were directed. For instance, the proliferative response of primed (2 × 13) F1 T cells to the copolymer L-glutamic and L-tyrosine (GT), controlled by a strain 13 Ir gene, was markedly inhibited in the presence of 2 anti-13 serum but not in the presence of 13 anti-2 serum, whereas 13 anti-2 selectively inhibited the response to DNP-GL, controlled by a strain 2 Ir gene, but not the response to GT. An extension of these studies was made possible by the availability of alloantisera which were directed to individual Ia specifities.[61] The selective inhibitory profile obtained in T cell proliferation assays in the presence of apparently monospecific anti-Ia sera allowed us to establish an association between individual Ia specifities and Ir genes. The following example should be given: the response to both GT and to 2,4-dinitrophenyl guinea pig albumin (DNP-GPA) is under control of strain 13 Ir genes. An antiserum directed to the Ia.1 specificity of strain 13 markedly inhibited the GT response, but did not affect significantly the response to DNP-GPA, whereas the DNP-GPA response was blocked only by anti-Ia.3, but not by anti-Ia.1.

The proposed genetic organization of the guinea pig I region represents a synopsis of serological, biochemical, and functional studies and is shown in Figure 5. The homology of different Ia antigens is based on the assumption that the three types of Ia molecules, which are distinguished by their subunit structure and molecular weight, are encoded by three distinct, independent loci representing three subregions of the guinea pig I region. Thus, Ia.2 of strain 2 and Ia.3,5 of strain 13 are both borne on molecule with noncovalently linked subunits of identical molecular weight and are, therefore, assigned to the same individual subregion. The available data do not yet allow an assignment of the Ia.6 specificity of strain 2 to a given subregion. Ia.5 and Ia.6 are considered as analogues of the public specificities of murine Ia antigens. The assignment of Ir genes to the three subregions is based on the ability of alloantisera directed to individual Ia determinants to specifically block T cell responses controlled by the Ir gene and on the association of the Ir gene with a given Ia determinant in an outbred population.

The I region of the guinea pig is probably located outside of the B and S regions and the genetic region bearing the genes for complement components is also likely to be outside of the B and S regions.[62] A crossover was detected in breeding studies by analyzing allotypic electrophoretic variants of C4. The C4 allotype segregated with the Ia antigens indicating that the complement region maps either between the I region and the B/S regions or, alternatively, outside of the I region. There is no indication for the existence of a region coding for T cell specific Ia antigens corresponding to the I-J region of the mouse. In addition, in the guinea pig, no regulatory T cell factors with helper or suppressor activity and bearing I region coded determinants have as yet been described.

V. EXPRESSION OF Ia ANTIGENS ON LYMPHOCYTES AND MACROPHAGES: ITS FUNCTIONAL SIGNIFICANCE AND ROLE IN CELLULAR INTERACTION

The inhibition of T lymphocyte proliferation in vitro by anti-Ia alloantisera initially suggested that the T cell was the site for the inhibitory effect.[60] This assumption, however, had to be reconsidered after the crucial role of the macrophage in the initiation

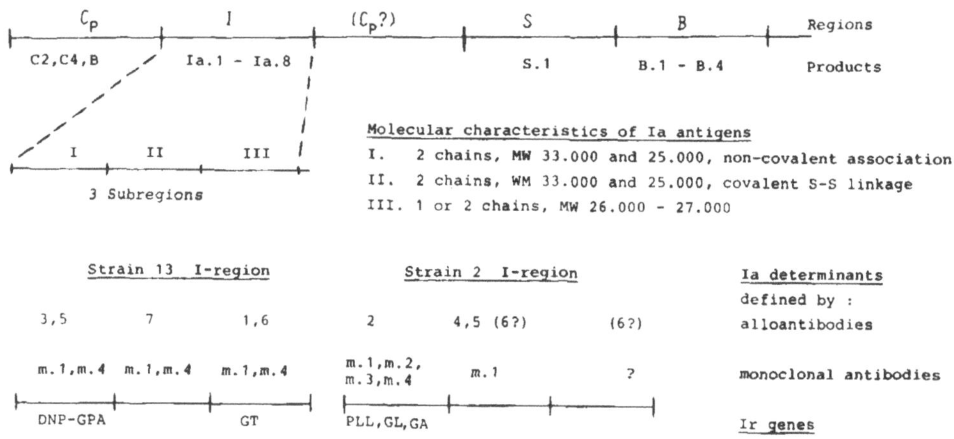

Figure 5. The genetic organization of the guinea pig major histocompatibility complex and of the I-region of strain 2 and strain 13 guinea pigs. The B- and S-regions on the chromosome code for histocompatibility antigens resembling the murine D- of K-region products. The Cp region controls the expression of the Complement Proteins C4, C2, and factor B; allelic genes for each component determine electrophoretic variants.[62] The chromosomal order of Cp-region is uncertain, but it appears to be located outside of the B- and S-regions. The individual Ir-specificities are assigned to individual I-subregions on the basis of their molecular characteristics. The determinants recognized by the monoclonal antibodies 22C4, 25E3, 25E11, or 27E7 are designated Ia.m1, Ia.m2, Ia.m3 or Ia.m4, respectively.

of the antigen-induced T cell proliferation was recognized.[63–65] Thus, it was possible that anti-Ia sera might exert their inhibitory effects by acting on the macrophage or, alternatively, on both the T cell and macrophage. Evaluation of the histocompatibility requirements for T lymphocyte-macrophage interaction revealed that only syngeneic antigen-pulsed macrophages were able to induce efficient T cell activation. In contrast, antigen-pulsed allogeneic macrophages failed to induce significant T cells stimulation. (2 × 13 F_1 T cells could be stimulated by either pulsed strain 2 or strain 13 parental macrophages.[63] The experiments described above measured the effect of the continuous presence of anti Ia sera on the T lymphocyte proliferation in a experimental system which did not allow one to distinguish the site of action of anti-Ia sera.

It was then demonstrated that the functional expression of Ir gene products at the macrophage level might be crucial. (2 × 13) F_1 lymphocytes could only be activated efficiently by antigen-pulsed macrophages of the responder type. Pulsed macrophages from the nonresponder parental strain failed to induce efficient T cell proliferation. For example, primed F_1 T cells were activated by DNP-GL pulsed strain 2 and F_1 macrophages, whereas GL-pulsed nonresponder strain 13 macrophages were not capable of activating F_1 lymphocytes. This result suggested that responder type macrophages express the required Ir gene product and that nonresponder macrophages are missing the Ir gene product necessary for efficient antigen processing or presentation.[64]

In a similar series of experiments, Ir gene function at the B cell level was analyzed.[66] Primed (2 × 13) F_1 cells were positively selected by culture with GL-pulsed, parental macrophages of the responder strain 2. These selected T cells functioned efficiently as T helper cells for the antigen, TNP-GL, only if cultured in the presence of hapten primed B cells of strain 2, but not of strain 13 origin. Thus, the functional expression of Ir gene products at the B cell level may also be required for efficient T helper cell-B cell interaction and antibody production.

Another approach to evaluate the functional expression of Ia antigens on immunocompetent cells involved negative selection utilizing anti-Ia serum and complement for elimination of Ia bearing subpopulations.[45,67] T cells or macrophages were first pretreated with anti-Ia sera and complement and the cells remaining alive were tested for their immunological functions. Two populations of T cells generated in vivo upon immunization could be distinguished by their differential susceptibility to lysis. The first subpopulation was lysed by anti-Ia sera and complement. The majority of primed T cells proliferating in vitro or producing migration inhibition factor (MIF) upon stimulation with antigen- or mitogen-pulsed macrophages belonged to this group. A second population was not susceptible to the lytic effects of the anti-Ia sera and complement and included the primed T cells which function as helper cells in the induction of an antibody response in vitro and the T cells responding to alloantigen in the mixed leukocyte reaction.[67]

The expression of Ia antigens on a minor, but functionally important, subpopulation of macrophages was recognized in the course of these studies. Only 15 to 25% of oil-induced macrophages were lysed by anti-Ia serum and complement. However, the cells remaining alive after treatment had a markedly reduced ability to function as antigen presentating cells or to act as stimulator cells in a mixed leukocyte reaction. Thus, the expression of Ia antigens on subpopulations of macrophages was required for efficient T cell activation induced by soluble protein antigens or by alloantigens.[45]

The development of an in vitro priming system allowed further crucial experiments on the localization of the cellular site of action of anti-Ia sera in the inhibition of antigen induced T cell proliferation.[68] Priming of nonimmune strain 13 T lymphocytes was obtained by co-culturing them for a 7-day period with antigen-pulsed or TNP-modified strain 13 macrophages. After an additional secondary culture of 3 days with fresh, pulsed syngeneic strain 13 or allogeneic strain 2 macrophages, the antigen-specific T cell proliferative response was determined. Only pulsed macrophages of the strain used in the priming culture were able to stimulate T cells efficiently in the secondary culture. In addition, in this system it was demonstrated that T cells could be primed with both syngeneic and allogeneic macrophages. A specific sensitization to antigen associated with allogeneic macrophages could be demonstrated after eliminating alloreactive T cells by negative selection with bromodeoxyuridine (BUdR) and light treatment.[69] The surviving T cells were then primed with allogeneic, antigen-pulsed macrophages and restimulated with pulsed allogeneic or syngeneic macrophages. Efficient T cell activation was only obtained in the second culture by restimulation with the allogeneic macrophages used for the initial sensitization.

This experimental system allowed one to determine whether either the macrophage, the T cell, or both are the cellular site of action of anti-Ia sera mediated inhibition of T cell proliferation.[70] Strain 13 T cells were depleted of alloreactive cells by BUdR and light treatment. The remaining cells were primed with TNP-modified strain 2 macrophages and responded in the second culture specifically to TNP-modified allogeneic strain 2 macrophages, but not syngeneic TNP-modified strain 13 macrophages. In the presence of anti-Ia.2,4 serum in the secondary culture, i.e., an alloantiserum directed solely against the stimulator macrophages, but not to the responder T cell, the strain 13 T lymphocyte proliferative response was inhibited. Anti-Ia.1,3,7 i.e., and antiserum directed exclusively against the responding strain 13 T cells had no inhibitory effect.

In summary, these data confirm that the target of anti-Ia sera in the inhibition of T cell activation is the antigen or alloantigen presenting macrophage, but not the responding T cell and suggest that the functional expression of Ir gene products is at the macrophage level. In addition, these findings demonstrate that the genetic restriction of the macrophage-T lymphocyte interaction is imposed by the histocompatibility type of the macrophage encountered in the initial sensitization and argue against the original hypothesis of the requirement for Ia homology between T cell and macrophage. The data are in agreement with the "complex antigenic determinant model" postulating that T cells recognize antigen in association with MHC products.[68,69]

VI. ANALYSIS OF Ia ANTIGEN/Ir GENE FUNCTION WITH MONOCLONAL ANTIBODIES

A. Alloantibodies Vs. Monoclonal Antibodies

One of the major tools in the analysis of Ia/Ir gene function was the use of alloantibodies in various in vitro systems. In particular, alloantibodies were utilized to study the role of Ia antigens in cellular interaction and to study Ir genes expressed on macrophages or T cells. Data obtained with alloantisera in a number of independent experimental studies, described in the previous section, suggested that Ia antigens might represent the actual Ir gene product as defined functionally. This assumption was supported by the fact that, despite intensive research for genetic recombinations within the MHC, no crossover was found which allowed one to distinguish the loci for Ir genes from the loci coding for Ia antigens. Another major question was whether the alloantigens with stimulatory capacity in the MLR were identical to the Ia antigens or the Ir gene products. The failure to distinguish by genetic studies the loci for these three traits was one argument in favor of the assumption that the Ia antigens, the Ir gene products, and the MLR stimulating alloantigens are all encoded by the same locus. The major support for this hypothesis was the ability of antisera directed to Ia antigens to specifically inhibit the activation of primed T cells by antigens, the response to which is controlled by Ir genes and to inhibit the MLR stimulatory capacity of Ia-bearing macrophages. However, the objection remained that the apparently monospecific antisera to Ia antigens actually contained antibodies to products of genes closely linked to the genes coding for Ia antigens. The specific inhibitory effects of the alloantisera might therefore be mediated by antibodies detecting undefined products of one or several loci closely linked but distinct from the Ia loci.

Monoclonal antibodies to Ia antigens have allowed us to resolve the question whether the inhibition of Ir gene controlled funtions by anti-Ia sera is actually mediated by antibodies to Ia antigens. The monoclonal nature of these reagents rules out the objection that antibodies to a postulated closely linked Ir gene product are perhaps responsible for the findings in functional tests. An additional advantage of monoclonal anti-Ia antibodies in functional studies is the ability to study the contribution of individual determinants of Ia molecules on Ia/Ir gene function. In particular the xenogeneic, monoclonal antibodies allow one to study functional properties of nonpolymorphic regions of Ia molecules against which alloantibodies cannot be raised.

The four monoclonal anti-Ia antibodies described in Section II.D. were, therefore, used in several in vitro T cell proliferation assays.[40] Antigen or mitogen induced T cell proliferation was measured by culturing adherence-column purified peritoneal exudate T cells from immunized animals with antigen- or mitogen-pulsed, mitomycin C-treated macrophages for 4 days in the continuous presence of the monoclonal antibodies. Ascitic fluid served as a source of antibody; ascitic fluid containing nonrelevant antibodies was used as a control in order to exclude nonspecific effects. T cell proliferation was measured by incorporation of tritiated thymidine. Similarly, inhibition of the MLR was tested by culturing purified lymph node T cells (LNL) with allogeneic or syngeneic, mitomycin C-treated macrophages in the presence of the monoclonal antibodies.

B. Inhibitory Effect of Monoclonal Anti-Ia Antibodies on the MLR

All four monoclonal anti-Ia antibodies react with strain 2 Ia antigens. In MLR cultures using strain 2 macrophages as stimulator cells and strain 13 LNL as responder cells, all four antibodies had an inhibitory effect. Antibodies 25E3 and 25E11, specific for strain 2 Ia antigens had no inhibitory effect on the MLR between strain 2 responder cells and strain 13 stimulator cells. Only antibodies 22C4 and 27E7 inhibited the MLR when strain 13 macrophages were used as stimulator cells. These two antibodies are directed against common determinants shared between Ia antigens of strain 2 and strain 13.

Proliferation in the presence of the monoclonal antibodies was usually reduced to about 30 to 40% of the control; a final dilution of 1/100 of the ascitic fluid in the cultures was used. The inhibitory effect was rapidly lost upon further dilution and was no longer present at a dilution of 1/800. This result is in contrast to the inhibitory effect on antigen induced proliferation where inhibition was observed at much higher dilutions (see below). It was the general experience in the MLR experiments that even at the highest concentrations of the monoclonals used, the maximal inhibition seen was less than that obtained with alloantisera. In addition, combinations of the different monoclonal antibodies failed to produce any additive inhibitory effects above the inhibition caused by an individual component of the mixture. Thus, despite the fact that the monoclonal antibodies were directed to different Ia determinants, alloantisera are definitely more efficient, indicating that multiple determinants contribute to MLR stimulation and that reagents with restricted specificity are, therefore, less efficient.

The fact that monoclonal antibodies to Ia antigens inhibit the MLR indicate that Ia antigens themselves are the stimulatory alloantigens. Expression of Ia antigens on the stimulator cells is required for stimulation and is the site of action of the anti-Ia antibodies because antibodies 25E3 and 25E11 inhibit the MLR only when they are directed against the stimulator strain 2 macrophage. No inhibition is present if the antibodies are directed against the responding strain 2 T cell, despite the demonstration of Ia expression by the T cells (see Section III.B.). In contrast, the antibodies 22C4 and 27E7 directed to common determinants shared between the two strains inhibit the MLR in both directions, presumably because in each case they are directed against the Ia antigens of the stimulator macrophages. The results obtained with the monoclonal reagents, therefore, present an unequivocal confirmation of the conclusions derived from experiments with alloantisera in several different species, that anti-Ia antibodies directed to stimulator cells, but not to responder T cells, inhibit the alloantigen induced response in the MLR.[71–73]

C. Selectivity of Inhibition in Antigen or Mitogen Induced T Cell Proliferation

The monoclonal anti-Ia antibodies were tested for their inhibitory effect on the antigen or mitogen induced T cell proliferative response of strain 2, strain 13 or (2 × 13) F_1 T cells. Macrophages pulsed with the antigens GT, GL, OVA, PPD, or the mitogen PHA were used. The cultures were performed in the continuous presence of the antibodies. Alloantisera in this system cause a marked inhibition of the response both to antigens, the response to which is controlled by Ir genes, and to antigens, the response to which is not under unigenic control.[74] The four monoclonal antibodies also inhibited the proliferative T cell response to antigens or to mitogens. However, the inhibitory profile obtained varied for each antibody and depended on the strains and the antigens which were used. In Figure 6, the dose-dependency and the selectivity of inhibition is demonstrated for antibodies 22C4 and 25E3. Primed (2 × 13) F_1 T cells were cultured with antigen- or mitogen-pulsed macrophages in the presence of increasing concentrations of antibody. A marked concentration-dependent inhibition of the GT- and PPD-induced proliferative response was obtained with antibody 22C4 up to a dilution of 1/8000. Maximum inhibition was about 60 to 70% as compared to the corresponding cultures containing control-ascites. In contrast, the response to GL or the mitogen PHA was not affected even at high concentrations of antibody 22C4.

Antibody 25E3 showed a different pattern of inhibition (Figure 6). The response to the antigen GL, which is controlled by an Ir gene of strain 2, was selectively inhibited,

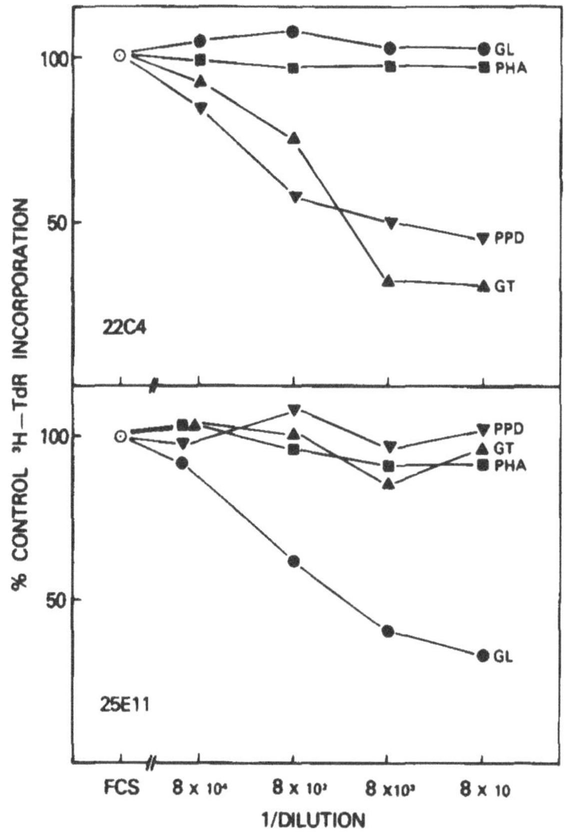

Figure 6. Inhibition of the proliferative response of (2 × 13) F_1 PEL to antigen- or mitogen-pulsed PEC. F_1 PEL (2 × 10^5) were cultured with antigen- or mitogen-pulsed F_1 PEC (1 × 10^5) in the presence of the indicated dilution of 22C4, 25E11, or the control ascites. Results are expressed as the percent of control ^{3}H-TdR incorporation:

$$= 100 \times \frac{\Delta \text{ cpm in anti-Ia ascites}}{\Delta \text{ cpm in control ascites}}$$

(From Berger, R. and Shevach, E. M., *J. Exp. Med.*, 152, 1011, 1980. With Permission.)

whereas the responses to PPD, GT, and PHA remained virtually unaffected. Table 3 summarizes the results obtained with the four antibodies in proliferation experiments with (2 × 13) F_1 T cells at a final ascites concentration of 1/800 in the cultures. Antibodies 25E3 and 25E11 specifically inhibited the response to GL-pulsed macrophages, but had no effect on other antigens or mitogens. Antibody 27E7 inhibited the response to all antigens tested as well as the mitogen PHA and differed, therefore, from antibody 22C4 which produced a marked inhibition of GT and PPD but not GL.

A corresponding inhibition pattern was obtained in the parental strains. Antibody 22C4 inhibited the GT response of strain 13 and the PPD response of both strain 2 and strain 13. Antibodies 25E3 and 25E11 only had a significant inhibitory effect on the GL response of strain 2, but did not affect the response to other antigens or the response of strain 13 T cells. Antibody 27E7 inhibited the antigen and mitogen responses of both parental strains to the same extent as that of (2 × 13) F_1 cells.

Table 3
EFFECT OF MONOCLONAL ANTI-Ia ANTIBODIES ON THE PROLIFERATIVE RESPONSE OF (2 × 13) F_1 PEL

	Stimulus			
	GL	GT	PPD	PHA
		Δ cpm		
22C4	39,300	5,749	44,031	62,440
25E3	14,011	12,339	82,398	54,577
25E11	10,834	13,867	109,836	57,321
27E7	5,063	6,908	48,438	29,867
Control ascites	32,507	15,058	95,237	65,438

Note: (2 × 13) F_1 PEL (2 × 10^5) were cultured for 3 days with antigen- or mitogen-pulsed syngeneic PEC (1 × 10^5) in the presence of a 1/800 dilution of the anti-Ia or the control ascites. Results are expressed as Δ cpm (^{3}H)TdR incorporation. Underlined results represent significant inhibition.

From Burger, R. and Shevach, E. M., *J. Exp. Med.*, 152, 1011, 1980. With permission.

D. Do Different Monoclonal Anti-Ia Antibodies Distinguish Functionally Different Regions of Ia Molecules Involved in Antigen Presentation?

The four monoclonal antibodies obviously differ in their inhibitory capacity from each other and from conventional antisera. Conventional alloantisera directed against multiple Ia determinants of either strain 2 or strain 13 inhibit the T cell response to a variety of antigens.[74] Antibody 27E7 resembles, in its broad inhibitory capacity, 2 anti-13 or 13 anti-2 alloantisera and is the only antibody which also inhibited the PHA response. Despite its uniform inhibitory effect, this antibody is directed only to the Ia.2 bearing molecule of the strain 2 I region, but not the Ia.4,5 bearing molecule. It reacts with all Ia antigens of strain 13. Antibodies 25E3 and 25E11 appeared to be directed to the alloantigenic determinant of the Ia.2 molecule of strain 2 on the basis of the serological assays. These antibodies specifically inhibited the response of strain 2 and (2 × 13) F_1 T cells to GL. Thus, these two monoclonal antibodies also functionally behave like an alloantiserum monospecific for Ia.2, which selectively blocks the GL response controlled by an Ia.2 linked Ir gene.

The reactivity profile exhibited by antibody 22C4 does not fit into a conventional pattern. This antibody in serological and immunochemical tests recognizes a common determinant, present on all Ia molecules of both strain 2 and strain 13. Despite this broad serological reactivity, a highly selective inhibitory effect was found. 22C4 clearly reacts in serological tests and in immunoprecipitation with the Ia molecule bearing the Ia.2 determinant known to be associated with the response to GL. Nevertheless, it fails to inhibit the GL response of strain 2 or (2 × 13) F_1 cells. Therefore, the determinant recognized by 22C4 on the Ia.2 bearing molecule must be functionally and spatially distinct from the antigenic determinant of the Ia.2 molecule which participates in the presentation of GL. Paradoxically, antibody 22C4 had a marked inhibitory effect on the response of strain 13 or (2 × 13) F_1 T cells to GT. The response to GT is controlled by an Ir gene present in strain 13, but absent in strain 2. The antibody was raised, however, against cells derived from the GT nonresponder strain 2. Thus, at least part of the molecular structures present in strain 13 on the Ia.1,6 molecule, and involved in presentation of GT, must also be present in the nonresponder strain 2. The corresponding molecule present in strain 2 lacks, however, the functionally critical moieties and is not able to confer responder status to strain 2. In the murine system, gene complementation is required for the generation of an immune response to a number of antigens.[75,76] A similar phenomenon might exist in the guinea pig. Antibody 22C4 might be directed to a determinant present on 1 of 2 complementary Ia chains which are both necessary to mount an immune response to GT and are both present in strain 13. Strain 2 might possess only the 1 chain which bears the determinant recognized by 22C4.

The fact that monoclonal anti-Ia antibodies specifically inhibit the activation of T lymphocytes by antigens (the response to which is controlled by Ir-genes) strongly suggests that the Ia antigens themselves are the products of Ir genes. The monoclonal nature of the antibodies rules out the possibility that the inhibitory effect of anti-Ia sera on T cell proliferation is mediated by antibodies directed to unique determinants of Ir gene products.

The cellular site of action of anti-Ia antibodies was localized at the macrophage level in a number of studies.[70,77] In view of the expression of Ia antigens on T cells (see Section III.B.), some of the inhibitory effects might result from an effect on the T cell. However, the specific and selective inhibition of T cell proliferation by the monoclonal antibodies would require a participation of Ia antigens at the T cell surface for antigen-specific recognition. Most evidence argues against this assumption.[76]

There are a number of different mechanisms which might explain the inhibition of T cell activation by anti-Ia antibodies. A generalized inhibitory effect, as obtained with alloantisera against multiple specificities or with antibody 27E7, could result from physical alteration of Ia molecules in the cell membrane. Binding of the antibodies may induce redistribution of Ia antigens on the cell surface resulting eventually in a biologically ineffective sterical arrangement in the membrane. Capping, followed by shedding or by internalization, might be the consequence of cross-linking of Ia antigens in the cell membrane. Alternatively, binding of antibodies could lead to configurational changes of the Ia molecules with a concomitant loss of biological activity in antigen presentation or recognition.

The antigen-specific inhibition mediated by the other monoclonal antibodies suggests a more specific function of Ia molecules and a possible mechanism of the specificity of Ir gene function. Two distinct hypotheses have been proposed to explain the specificity of the Ir gene controlled immune response.[78,79] The first postulates a specific and selective interaction of the antigen or of antigen fragments with Ia molecules expressed by the antigen presenting macrophages.[79,80] A broad polymorphism of I region products would, therefore, be required for generation of distinct molecular configurations of Ia antigens which would specifically interact with antigen fragments to form the immunogenic complex recognized by the antigen specific T cell.

The selective inhibitory profile of the monoclonal anti-Ia antibodies is consistent with this hypothesis. Thus, monoclonal antibodies might discriminate between different parts of Ia molecules which have different functional roles in antigen presentation. Certain parts of an Ia molecule might participate in the presentation of certain antigens or antigen fragments, but not of others. If the individual determinants on the Ia molecules are recognized by the monoclonal antibodies, efficient presentation of some antigens would be inhibited. Adjacent parts of the same Ia molecule, or other Ia molecules involved in presentation of distinct antigens, may still be able to express their biological functions.

Alternatively, our results are also compatible with theories of Ir gene function which postulate a selective T cell differentiation step in the thymus during which individual clones are restricted to specificities for certain antigens and selected Ia molecules.[81] Thus, the T cell has two receptors, one for self-Ia on macrophages and the other for conventional antigens and Ir genes function during T cell differentiation in the thymus to limit the combinations of anti-self Ia and anti-antigen receptors that can be expressed by a population of T cells. Monoclonal anti-Ia antibodies would inhibit T cell proliferation by masking the recognition of individual Ia epitopes seen by the anti-self component of the T cell receptor.

The validity of these hypotheses should be tested by analysis with a larger panel of monoclonal antibodies. Such a detailed evaluation might define the relationship between the obvious structural variability of Ia antigens as defined serologically and the corresponding biological or functional activity in antigen presentation. It might then be possible to correlate the presence of a given antigenic determinant on an Ia molecule with a specific immunologic function.

VII. CONCLUDING REMARKS

This article has attempted to summarize the recent studies on the serology, structure, and function of the Ia antigens of the guinea pig with focus on the work of our laboratory. Because of the limited number of inbred strains in this species, analysis is made more difficult than in the mouse. The efficient combination of serological and immunochemical methods with functional studies nevertheless allowed considerable progress to be made. Further application of monoclonal antibodies to Ia antigens, especially in combination with monoclonal T cell lines, should allow further elucidation of the precise mechanism of action and the cellular site of expression of Ia antigens and Ir genes.

REFERENCES

1. **Kantor, F. S., Ojeda, A., and Benacerraf, B.,** Studies on artificial antigens. I. Antigenicity of DNP-polylysine and DNP-copolymer of lysine and glutamic acid in guinea pigs, *J. Exp. Med.*, 117, 55, 1963.
2. **McDevitt, H. O. and Benacerraf, B.,** Genetic control of specific immune responses, *Adv. Immunol.*, 11, 31, 1969.
3. **Benacerraf, B. and McDevitt, H. O.,** The histocompatibility linked immune response genes, *Science*, 1975, 273, 1972.
4. **Benacerraf, B. and Katz, D. H.,** The histocompatibility-linked immune response genes, *Adv. Cancer Res.*, 21, 121, 1975.
5. **Shevach, E. M.,** The guinea pig I-region. A functional analysis of Ia-Ir association, *Springer Semin. Immunopathol.*, 1, 207, 1978.
6. **Schwartz, B. D. and Cullen, S. E.,** Chemical characteristics of Ia antigens, *Springer Semin. Immunopathol.*, 1, 85, 1978.
7. **Schwartz, B. D., Paul, W. E., and Shevach, E. M.,** Guinea pig Ia antigens: functional significance and chemical characterization, *Transplant. Rev.*, 30, 1974, 1976.
8. **Wright, S.,** The affects of inbreeding and cross-breeding on guinea pigs. I. Decline in vigor, *U.S. Dep. Agr. Bull.*, 1090, 1, 1922.
9. **Sato, W. and de Weck, A. L.,** Leukocyte typing in guinea pigs, *Z. Immunitatsforsch. Allerg. Klin. Immunol.*, 144, 49, 1972.
10. **Bluestein, H. G., Ellman, L., Green, I., and Benacerraf, B.,** Specific immune response genes of the guinea pig. III. Linkage of the GA and GT immune response genes to histocompatibility genotypes in inbred guinea pigs, *J. Exp. Med.*, 134, 1529, 1971.
11. **Ellman, L., Green, I., Martin, W. J., and Benacerraf, B.,** Linkage between the poly-L-lysine gene and the locus controlling the major histocompatibility antigens in strain 2 guinea pigs, *Proc. Natl. Acad. Sci. U.S.A.*, 66, 322, 1970.

12. **Shevach, E. M., Rosenstreich, D. L., and Green, I.,** The distribution of histocompatibility antigens on T and B cells in the guinea pig, *Transplantation,* 16, 126, 1973.
13. **Geczy, A. F., de Weck, A. L., Schwartz, B. D., and Shevach, E. M.,** The major histocompatibility complex of the guinea pig. I. Serologic and genetic studies, *J. Immunol.,* 115, 1704, 1975.
14. **Schwartz, B. D., Kask, A. M., Paul, W. E., and Shevach, E. M.,** Structural characteristics of the allonantigens determined by the major histocompatibility complex of the guinea pig, *J. Exp. Med.,* 143, 541, 1976.
15. **Bjorek, L. R., Cigen, B., Bergard, B., Low, B., and Bergard, I.,** Relationship between β2-microglobulin and alloantigens coded for by the major histocompatibility complexes of the rabbit and the guinea pig, *Scand. J. Immunol.,* 6, 1063, 1977.
16. **Schwartz, B. D., McMillan, M., Shevach, E. M., Hahn, Y., Rose, S. M., and Hood, L.,** Partial N-terminal amino acid sequences of guinea pig classic histocompatibility antigens, *J. Immunol.,* 125, 1055, 1980.
17. **Shevach, E. M., Geczy, A. F., de Weck, A., and Schwartz, B. D.,** The guinea pig I-region. III. Distribution of Ia antigens in inbred and partially inbred strains, *J. Immunogenetics,* 7, 229, 1980.
18. **Schwartz, B. D., Kask, A. M., Paul, W. E., Geczy, A. F., and Shevach, E. M.,** The guinea pig I region. I. A structural and genetic analysis, *J. Exp. Med.,* 146, 547, 1977.
19. **Altman, P. L. and Katz, D. D.,** Inbred and genetically defined strains of laboratory animals. Part 2. Hamster, guinea pig, rabbit and chicken, *Federation of American Societies for Experimental Biology,* 1979, chapter 4.
20. **Bitter-Suermann, D., Kronke, M., Hoffman, T., Geczy, A. F., and Hadding, U.,** Evidence for a linkage group of the polymorphic complement proteins C4, C2 and factor B related to the MHC of the guinea pig (GPLA), *J. Immunol.,* 120, 1765, 1978.
21. **Schwartz, B. D. and Nathenson, S. G.,** Isolation of H-2 alloantigens solubilized by the detergent NP-40, *J. Immunol.,* 107, 1363, 1971.
22. **Cullen, S. E., Freed, J. H., and Nathenson, S. G.,** Structural and serological properties of murine Ia alloantigens, *Transplant. Rev.,* 30, 236, 1976.
23. **Cullen, S. E. and Schwartz, B. D.,** An improved method for isolation of H-2 and Ia alloantigens using immunoprecipitation induced by protein A bearing staphylococcus, *J. Immunol.,* 117, 136, 1976.
24. **Cullen, S. E., Freed, J. H., Atkinson, P. M., and Nathenson, S. G.,** Evidence that protein determines Ia antigenic specificity, *Transplant. Proc.,* 8, 237, 1975.
25. **Vitetta, E. S., Klein, G., and Uhr, J. W.,** Partial characterization of Ia antigens from murine lymphoid cells, *Immunogenetics,* 1, 82, 1974.
26. **Clement, L. T., Kask, A. M., and Shevach, E. M.,** Rapid purification of detergent-solubilized Ia antigens by immunoabsorbent chromatography, *Immunochemistry,* 15, 393, 1978.
27. **Kessler, S. W.,** Rapid isolation of antigens from cells with a staphylococcal protein A antibody absorbent. Parameters of the interaction of antibody-antigen complexes with protein A, *J. Immunol.,* 115, 1617, 1975.
28. **Finkelman, F. E., Shevach, E. M., Vitetta, E. S., Green, I., and Paul, W. E.,** Guinea pig immune response related histocompatibility antigens. Partial characterization and distribution, *J. Exp. Med.,* 141, 27, 1975.
29. **Schwartz, B. D., Vitetta, E. S., and Cullen, S. E.,** Labeling characteristics and separation of Ia antigen subunits, *J. Immunol.,* 120, 671, 1978.
30. **Jones, P. P., Murphy, D. B., Hewgill, D., and McDevitt, H. O.,** Detection of a common polypeptide chain in I-A and I-E subregion immunoprecipitates, *Mol. Immunol.,* 16, 51, 1979.
31. **Waxdal, M. J., Basham, T. Y., Clement, L. T., Shevach, E. M., and Schwartz, B. D.,** Amino terminal sequence studies on the Ia antigens of the guinea pig, *Mol. Immunol.,* 16, 61, 1979.
32. **Sachs, D. H., Winn, H. J., and Russel, P. S.,** The immunologic response to xenografts. Recognition of mouse H-2 histocompatibility antigens by the rat, *J. Immunol.,* 107, 481, 1971.
33. **McKenzie, I. F. C. and Painter, M. E.,** The antibody response of rabbits to soluble mouse H-2 antigens, *J. Immunol.,* 108, 352, 1972.
34. **Ferrone, S., Natali, P. G., Hunter, A., Terasaki, P. I., and Reisfeld, R. A.,** Immunogenicity of soluble HL-A alloantigens, *J. Immunol.,* 108, 1718, 1972.
35. **Metzgar, R. S. and Miller, J. L.,** Production of HL-A-Specific antibodies in monkeys by immunization with soluble HL-A antigens, *Transplantation,* 13, 467, 1971.
36. **Kohler, G. and Milstein, C.,** Continuous culture of fused cells secreting antibody of predefined specificity, *Nature (London),* 256, 495, 1975.
37. **Kohler, G. and Milstein, C.,** Derivation of specific antibody-producing tissue culture and tumor lines by cell fusion, *Eur. J. Immunol.,* 6, 511, 1976.
38. **Clement, L. T., Kask, A. M., Schwartz, B. D., and Shevach, E. M.,** Analysis of guinea pig membrane proteins recognized by heterologous antilymphocyte sera. Specific recognition of Ia alloantigens, *Transplantation,* 27, 397, 1979.

39. **Burger, R., Clement, L., Schroer, J., Chiba, J., and Shevach, E. M.**, Monoclonal antibodies to guinea pig Ia antigens. I. Production, serologic and immunochemical characterization, *J. Immunol.*, 126, 32, 1981.
40. **Burger, R. and Shevach, E. M.**, Monoclonal antibodies to guinea pig Ia antigens. II. Effect on alloantigen-, antigen-, and mitogen-induced T lymphocyte proliferation in vitro, *J. Exp. Med.*, 152, 1011, 1980.
41. **Galfre, G., Howe, S. C., Milstein, C., Butcher, G. W., and Howard, J. C.**, Antibodies to major histocompatibility antigens produced by hybrid cell lines, *Nature (London)*, 266, 550, 1977.
42. **Burger, R., Scher, I., Sharrow, S., and Shevach, E. M.**, Non-activated guinea pig T lymphocytes and thymocytes express Ia antigens: FACS analysis with alloantibodies and monoclonal antibodies, in press.
43. **Forni, G., Rhim, J. S., Pickeral, S., Shevach, E. M., and Green, I.**, Antigeniticy of carcinogen and viral induced sarcomas in inbred and randombred guinea pigs, *J. Immunol.*, 115, 204, 1975.
44. **Shevach, E. M., Paul, W. E., and Green, I.**, Alloantiserum-induced inhibition of immune response gene product function. I. Cellular distribution of target antigens, *J. Exp. Med.*, 139, 661, 1974.
45. **Yamashita, U. and Shevach, E. M.**, The expression of Ia antigens on immunocompetent cells in the guinea pig. II. Ia antigens on macrophages, *J. Immunol.*, 119, 1584, 1977.
46. **Rogoff, T. M. and Lipsky, P. E.**, Accessory cell function of Ia bearing Kupffer cells, in *Abstracts 4*, Preud'Homme, I. L. and Hawken, V. A. L., Eds., Int. Congress of Immunology, Paris, Intl. Union of Immunol. Soc., 1980, 2.
47. **Stingl, G., Katz, S. I., Shevach, E. M., Wolff-Schreiner, E., and Green, I.**, Detection of Ia antigens on Langerhans cells in guinea pig skin, *J. Immunol.*, 120, 570, 1978.
48. **Stingl, G., Katz, S. I., Clement, L., Green, I., and Shevach, E. M.**, Immunologic function of Ia-bearing epidermal Langerhans cells, *J. Immunol.*, 121, 2005, 1978.
49. **Clement, L. T., Hu, C. P., Shevach, E. M., and Green, I.**, Immunologic characterization of a granulocytic leukemia of inbred strain 13 guinea pigs. Presence of Ia-positive myeloblasts, *Blood*, 55, 304, 1980.
50. **Schwartz, B. D., Gordon, D., and Thomas, D. W.**, Further chemical characterization of guinea pig Ia molecules derived from the three major classes of immunocompetent cells, *Mol. Immunol.*, 16, 43, 1979.
51. **Nagy, Z. A., Elliott, B. E., Nabholz, M., Krammer, P. H., and Pernis, B.**, Specific binding of alloantigens to T cells activated in the mixed lymphocyte reaction, *J. Exp. Med.*, 143, 648, 1976.
52. **Yu, D. T. Y., McCune, J. M., Fu, S. M., Winchester, R. J., and Kunkel, H. G.**, Two types of Ia positive T cells. Synthesis and exchange of Ia antigens, *J. Exp. Med.*, 152, 89s, 1980.
53. **Schwartz, B. D., Kask, A. M., Paul, W. E., and Shevach, E. M.**, Chemical differences between Ia antigens and subpopulations of guinea pig lymphocytes, *Fed. Proc., Fed. Am. Soc. Exp. Biol.*, 35, 858, 1976.
54. **David, C. S., Meo, T., McCormick, J., and Shreffler, D.**, Expression of individual Ia specificities on T and B cells. I. Studies with mitogen-induced blast cells, *J. Exp. Med.*, 143, 218, 1976.
55. **Ko, H. S., Fu, S. M., Winchester, R. J., Yu, D. T., and Kunkel, H. G.**, Ia determinants on stimulated human T lymphocytes. Occurrence on mitogen- and antigen-activated T cells, *J. Exp. Med.*, 150, 246, 1979.
56. **Reinherz, E. L., Kung, P. C., Pesando, J. M., Rik, J., Goldstein, G., and Schlossman, S. F.**, Ia determinants on human T-cell subsets defined by monoclonal antibody. Activation stimuli required for expression, *J. Exp. Med.*, 150, 1472, 1979.
57. **Geczy, A. F. and de Weck, A. L.**, The major histocompatibility complex of the guinea pig. III. Relationship between mixed leukocyte reactivity and serologically-defined phenotypes of the GPLA B locus and I region, *J. Immunogenet.*, 3, 191, 1976.
58. **Braley-Mullen, H., Sharp, G. C., and Kyriaskos, M.**, Autoimmune thyroditis in guinea pig. Association with histocompatibility type, *Immunogenetics*, 3, 205, 1976.
59. **Hyman, L. R., Stembey, A. D., Colvin, R. B., and Bernard, E. F.**, Immunopathogenesis of autoimmune tubulointerstitial nephritis. II. Role of an immune response gene linked to the major histocompatibility complex, *J. Immunol.*, 117, 1894, 1976.
60. **Shevach, E. M., Paul, W. E., and Green, I.**, Histocompatibility-linked immune response gene function in guinea pigs. Specific inhibition of antigen-induced lymphocyte proliferation by alloantisera, *J. Exp. Med.*, 136, 1207, 1972.
61. **Shevach, E. M., Lundquist, M., Geczy, A. F., and Schwartz, B. D.**, The guinea pig I region. II. Functional analysis, *J. Exp. Med.*, 146, 561, 1977.
62. **Kronke, M., Hadding, U., Geczy, A., de Weck, A., and Bitter-Suermann, D.**, Linkage of guinea pig Bf and C4 to the GPLA, *J. Immunol.*, 119, 2016, 1977.

63. **Rosenthal, A. S. and Shevach, E. M.,** The function of macrophages in antigen recognition by guinea pig T lymphocytes. I. Requirement for histocompatible macrophages and lymphocytes, *J. Exp, Med.*, 138, 1194, 1973.
64. **Shevach, E. M. and Rosenthal, A. S.,** The function of macrophages in antigen recognition by guinea pig T lymphocytes. II. Role of the macrophage in the regulation of genetic control of the immune response, *J. Exp. Med.*, 138, 1213, 1973.
65. **Waldron, J. A., Horn, R. G., and Rosenthal, A. S.,** Antigen induced proliferation of guinea pig lymphocytes in vitro: obligatory roles of macrophages in the recognition of antigen by immune lymphocytes, *J. Immunol.*, 111, 58, 1973.
66. **Yamashita, U. and Shevach, E. M.,** The histocompatibility restriction on macrophage T-helper cell interaction determine the histocompatibility restriction on T-helper cell B-cell interaction, *J. Exp. Med.*, 148, 1171, 1978.
67. **Yamashita, U. and Shevach, E. M.,** The expression of Ia antigens on immunocompetent cells in the guinea pig. I. The differential expression of Ia antigens on T cell subpopulations, *J. Immunol.*, 119, 1575, 1977.
68. **Thomas, D. W. and Shevach, E. M.,** Nature of the antigenic complex recognized by T lymphocytes. I. Analysis with an in vitro primary response to soluble protein antigens, *J. Exp. Med.*, 144, 1263, 1976.
69. **Thomas D. W. and Shevach, E. M.,** Nature of the antigenic complex recognized by T lymphocytes. Specific sensitization by antigens associated with allogeneic macrophages, *Proc. Natl. Acad. Sci. USA*, 74, 2104, 1977.
70. **Thomas D. W., Yamashita, U., and Shevach, E. M.,** Nature of the antigenic complex recognized by T lymphocytes. IV. Inhibition on antigen-specific T cell proliferation by antibodies to stimulator macrophage Ia antigens, *J. Immunol.*, 119, 223, 1977.
71. **Greineder, D. K., Shevach, E. M., and Rosenthal, A. S.,** Macrophage-lymphocyte interaction. III. Site of alloantiserum inhibition of T lymphocyte proliferation induced by allogeneic or aldehyde-bearing cells, *J. Immunol.*, 117, 1261, 1976.
72. **Greenberg, L. J., Reinsmoen, N., and Yunis, E. J.,** Dissociation of stimulation (MLR-S) and response (MLR-R) in mixed leukocyte culture by serum blocking factors, *Transplanation*, 16, 520, 1973.
73. **Meo, T., David, C. S., Rijubeek, A. M., Miggiano, V. C., and Shreffler, D. C.,** Inhibition of mouse MLR by anti-Ia sera, *Transplantation Proc.*, 7s1, 127, 1975.
74. **Ruhl, H. and Shevach, E. M.,** The effect of alloantisera on antigen-induced T cell proliferation, *J. Immunol.*, 115, 1493, 1975.
75. **Schwartz, R. H., Dorf, M. E., Benacerraf, B., and Paul, W. E.,** The requirement for two complementing Ir-GLØ immune response genes in the T-lymphocyte proliferation response to poly (Glu^{53}-Lys^{36}-Phe^{11}), *J. Exp. Med.*, 143, 897, 1976.
76. **McDevitt, H. O.,** The role of H-2 I region genes in regulation of the immune response, in *Immunology 80*, Fougerau, M. and Dausset, J. Eds., Academic Press, London, 1980, 503.
77. **Longo, D. and Schwartz, R. H.,** Gene complementation. Neither Ir-GLØ gene need be present in the proliferative T cell to generate an immune response to Poly (Glu^{55}-Lys^{36}-Phe^{9}), *J. Exp. Med.*, 151, 1452, 1980.
78. **Benacerraf, B.,** Genetic control of the specificity of T lymphocytes and their regulatory products, in *Immunology 80*, Fougerau, M. and Dausset, J. Eds., Academic Press, London, 1980, 419.
79. **Benacerraf, B.,** A hypothesis to relate the specificity of T lymphocytes and the activity of I region-specific Ir genes in macrophages and B lymphocytes, *J. Immunol.*, 120, 1809, 1978.
90. **Rosenthal, A. S., Barcinski, M. A., and Blake, T. J.,** Determinant selection is a macrophage dependent immune response gene function, *Nature (London)*, 267, 156, 1977.
81. **Von Boehmer, H., Haas, W., and Jerne, N. K.,** Major histocompatibility complex-linked immuneresponsiveness is acquired by lymphocytes of low-responder mice differentiating in thymus of high-responder mice, *Proc. Natl. Acad. Sci. USA*, 75, 2439, 1978.

Chapter 3

SEROLOGY AND GENETICS OF HUMAN Ia-ANTIGENS*

E. D. Albert

TABLE OF CONTENTS

*Supported by DFG A1 92/12 and SFB 37.

I. HISTORIC REMARKS

In man, the discovery of the "Ia" equivalents followed a different course than that in the mouse. In experimental animals, the I region was defined by immune response (IR) genes and the Ia antigens were found as a result of the attempt to define the gene products of the IR genes.[1] In man there is—even today—only a very limited amount of information available about HLA-linked IR genes. Important information was obtained in studies with mixed leukocyte cultures (MLC), where it is known from the mouse that the major determinants are coded for in the I region. By analogy it is generally assumed that the MLC determinants in man are also located in the immune response region. Typing for MLC determinants became possible through the use of homozygous typing cells.[2] This procedure led to the definition of six allelic specificities (HLA-Dw1, Dw2, Dw3, Dw4, Dw5, Dw6) by the Sixth International Histocompatibility Workshop 1975.[3] The locus coding for the major MLC determinants was designated as HLA-D. Since the procedure of typing for HLA-D with homozygous typing cells is rather cumbersome, time consuming, and not always easy to interpret, attempts were made to define MLC antigens by serologic means. Selected antisera, which inhibited the MLC reaction between unrelated, HLA-A,B,C identical individuals were found to contain antibodies, directed against antigens, which were—just like the murine Ia antigens—predominantly expressed on B cells and absent from T cells and platelets.[4,5] This has prompted the development of a procedure for B cell enrichment of lymphocyte suspensions, which then could be tested in a slight modification of the standard microlymphocytotoxicity assay.[6] Antisera containing anti-HLA-A,B,C antibodies in addition to the "B cell-specific" antibodies can be purified by absorption with two volumes of packed platelets. Some of the sera thus prepared showed a very close correlation with the HLA-D antigens determined by MLC using homozygous typing cells.[5] The Seventh International Histocompatibility Workshop resulted in the definition of eight "B cell specific" HLA antigens, each of which exhibits a strong correlation with one HLA-Dw antigen as defined by MLC. Since it was not clear whether serology and MLC are detecting the same antigens, the designation "D-related" (=DR) was chosen by the WHO nomenclature committee.[7]

II. TECHNICAL ASPECTS OF DR TYPING

A. Selection of Sera

Antisera containing anti-HLA-DR antibodies can be obtained from multiparous women and from patients who have been repeatedly transfused. Patients who have rejected a transplant also produce anti-HLA-DR antibodies with high frequency. Such sera are, however, rarely obtained in satisfactory quantity. Screening for B cell antibodies may be performed by testing antisera against separated B cells and separated T cells. In the early phases of HLA-DR serology, antisera were selected which reacted only with B cells and not with T cells. However such sera have been generally rather weak and undependable. It is the general experience of DR serologists that antisera which contain high titers of anti-HLA-A,B, and C antibodies frequently also include strong anti-DR antibodies. Therefore the screening policy shifted to using highly reactive anti-HLA-A,B,C antisera for screening.

B. Absorption Procedures

In order to remove the anti-HLA-A,B,C antibodies, sera can be absorbed with platelets from a donor who possesses the HLA-A,B,C antigens against which the antibodies are directed. Since such donors are generally not available, it has become the practice to absorb with a large volume of platelets pooled from at least 20 different donors. After absorption with two volumes of packed-platelets, the anti-HLA-A,B,C antibodies are in most cases totally removed from the antiserum. It is, however, necessary to investigate absorbed sera by testing them against several different T cell suspensions in order to be certain that all anti-HLA-A,B,C antibody is indeed absorbed out.

C. B Cell Enrichment of the Cell Suspension

The quality of the HLA-DR typing is largely dependent on the purity of the B cell suspension prepared from peripheral lymphocytes. Enrichment for B cells can be principally obtained through two different ways:

1. Rosetting with sheep red blood cells: a normal lymphocyte suspension is mixed with a suspension of sheep red blood cells. The T cells form rosettes with sheep red blood cells which can be stabilized through the treatment of these red blood cells with either neuraminidase or with AET. The rosettes formed by T cells and sheep red blood cells can be separated from the rest of the lymphocytes by ficoll-gradient-centrifugation. This method was introduced by van Rood.[6]
2. Adherence of B cells to nylon wool: B cells have a tendency to adhere to plastic surfaces and nylon wool. This property has been successfully used for the separation of B cells from T cells. A normal lymphocyte suspension is added onto a straw which is stuffed with washed nylon wool. After 30 min incubation at 37°C the non-adherent T cells are washed out by simply adding Hank's balanced salt solution to the straw. After thorough washing, the straw with the nylon wool is squeezed to release the B cells which are adherent to the nylon wool. This method can yield excellent cell preparations provided one has a large number of lymphocytes to begin with.

D. Complement

It has been the experience of many laboratories that the use of a special complement for DR testing is very important. Not each batch of complement which is perfect for regular HLA-A,B,C typing may be used for HLA-DR typing. Therefore it is necessary to perform a careful selection of the complement. The nature of the variation introduced into DR typing by the complement is not known at present.

E. Cytotoxicity Test

Since many B cell antisera are rather weak, it has been decided to perform a cytotoxicity test with prolonged incubation time. The suspension of B cells is incubated for 1 hr with the antisera and for 2 additional hours with complement. The incubation of cells with serum is performed at 37°C and the incubation with complement at room temperature (20°C).

F. Two-Color Fluorescence

An alternative method of B cell typing was described by van Rood et al. 1976.[8] It utilizes a fluorescence doublestaining technique in which the B cells are marked with FITC-coupled anti-immunoglobulin and cytotoxicity is demonstrated by a staining of killed cells with ethidium bromide. Thus B cells are marked with a green membrane staining while T cells lack this membrane staining. Killed cells take up ethidium bromide and appear with a bright red fluorescence. In this way, unseparated B and T cells suspensions can be tested and read using a filter combination which permits red and green fluorescence to be observed simultaneously.

G. Serological Pitfalls

A number of serological problems can cause considerable variations of DR typing results. One important variable is the quality of the antisera. It is clear that strong antisera tend to give better results and that weak antisera generally have a low degree of reproducibility.

The second most important variable is the quality of the B cell preparations. If the B cell enrichment does not go beyond 50% B cells in the final cell suspension, results may be quite variable. This problem will be recognized by generally weak reactions. Since B cells express a larger amount of HLA-A,B,C antigens than T cells, it is possible that weak anti-HLA,A,B,C antibodies in the sera used for DR typing may cause false positive, B cell-specific reactions. This can only be avoided through absorption with washed platelets.

Another variable is the possible mixture of cytotoxic antibodies which are directed against non-HLA determinants. Some antisera contain antibodies with an optimum of reactivity at low temperatures. Such "cold" antibodies, which also have auto-antibody nature, have a preference for B cells. They may be directed at the immunoglobulin on the surface of B cells. Another frequent contamination of HLA-DR antisera are antibodies directed against Lewis blood group. A contamination may be detected by careful monitoring of the Lewis blood types and eliminated by the addition of blood-group-substance to the antisera.

III. SEROLOGY AND GENETICS OF HLA-DR ANTIGENS

A. Definition of DR Antigens

1. Clustering Procedure Using the 2 × 2 Table

The definition of DR antigenic specificities follows the classical pattern of HLA-A,B,C serology: a large number of antisera are tested against a large panel of cell samples and the reaction patterns of single sera are compared in a 2 × 2 analysis. The degree of similarity is measured by the correlation-coefficient r and the level of significance is determined by the chi-square-value. Sera with similar reaction patterns are selected to form groups in which each member of the group reacts very similarly with each other member. In a good cluster of antisera, each individual pair should have a correlation coefficient of 0,8 or better.[9] After the establishment of such a group of highly positively associated antisera, it is possible to use this group of sera for the typing of the cell panel. A cell is considered positive for the antigen defined by the cluster if at least four out of five sera react positively with that cell. The distribution of the newly defined antigen in the test panel is then compared with that of other already established antigens in the same panel by means of a 2 × 2 table. If two antigens show a significant overrepresentation of discordant results in the 2 × 2 table (=negative correlation) this can be taken as evidence for a possible allelism of these two antigens, provided the panel consists of randomly selected unrelated individuals. Proof of allelism, however, must be obtained by extensive family data: the two antigens in question must never be found segregating together on the same haplotype.

If 2 antigens show a 2 × 2 table with positive correlation, it is necessary to recognize inclusion phenomena. The inclusion phenomenon is characterized by the finding that the discordant results of the 2 × 2 table is 0, or almost 0, while the other discordant field may contain a large number. Inclusion phenomena are rather frequent in HLA serology and indicate that both specificities are coded for at the same locus and that one specificity is supertypic to the other.

If 2 antigens show a positive correlation with significant numbers in both discordant fields of the 2 × 2 table, this may indicate that the 2 antigens belong to 2 different loci and that the 2 antigens show linkage disequilibrium in the unrelated population.

Since most HLA-DR antisera must be considered potentially multispecific (= containing many different antibody populations) typing for DR antigens is only safe if several antisera with the same specificity are used in the typing.

2. Family Clustering Analysis

An alternative method for the definition of DR antigens has been developed in family studies.[10] A panel of completely unselected anti-HLA-DR antisera is tested against a large group of families. For each family these sera are ordered according to their reaction patterns marking the four HLA haplotypes in the family. The group of antisera that reacts with the paternal haplotype A is further investigated in other families where the sera react with one of the parents. In the second family, sera which do not follow the group in their reaction pattern will be eliminated. If this procedure is continued for a number of families, a group of antisera will be established which defines one antigenic specificity. This method has been successfully applied to the data of the Seventh International Histocompatibility Workshop and the resulting groups have been virtually identical to those defined in the analysis based on a cluster formation in the unrelated population as described above. The family analysis method has the advantage that it favors the definition of subtypic rather than of supertypic specificities.

B. Description of Currently Recognized DR Antigens[11]

DR1 is a relatively well-defined antigen without serological evidence for heterogeneity. It shows cross-reactive patterns with DR2 and DRw6 as well as with DR5. DR1 is in the Caucasian population fairly strongly associated with Bw35 and more weakly so with B27.

DR2 is very well defined by many good antisera and it does not show any serological evidence for a possible split. The main cross-reactivity pattern of DR2 is with DR1 and DRw6. DR2 displays a strong linkage disequilibrium with HLA-B7 and it is known for its association with multiple sclerosis.

The antigen DR3 is clearly defined in Caucasians and in Negroes, and almost absent in the Oriental population. In spite of some evidence for heterogeneity of Dw3, there is no serological hint for any splitting. DR3 cross-reacts strongly with DRw6 and between DR3 and B8. DR3 shows a disease association with juvenile diabetes mellitus, celiac disease and a number of organ specific autoimmune diseases.

After initial difficulties in the Seventh Workshop, DR4 has been quite well defined in the Eighth Histocompatibility Workshop. After the splitting-off of DRw9 and DRw10, there is no further indication for heterogeneity of DR4. There are two groups of cross-reactions, one is with DR7 and the other cross-reactive pattern is with DR5 and DR6. DR4 shows only a relatively weak association with B15. DR4 is known for its involvement in disease association with juvenile onset diabetes and with rheumatoid arthritis.

DR5 is an antigen with some evidence of heterogeneity both from serological and from cellular work. This antigen displays three different types of cross-reactive patterns: one with DR1, another with DRw6, and a third one with DRw8. There is no strong association of DR5 with B locus antigens.

The antigen DRw6 has been the continuing nightmare of HLA-DR serologists. Historically its definition depended on the reaction with two groups of antisera: one group which contains anti-DR3 and DRw6 activity, and a second group which contains anti-DR1, DR2, and DRw6 activity.[9] Due to the emphasis on the supertypic group system MT1,2,3 during the Eighth Histocompatibility Workshop, this definition of DRw6 was left and a major confusion resulted. Presently it appears that there are at least 3

Table 1
GENE FREQUENCIES OF OFFICIALLY RECOGNIZED HLA-DR-ANTIGENS IN THREE DIFFERENT POPULATIONS[34]

Antigen	Europ. Canc.	Negroes	Japanese	Linkage disequilibrium	Disease association[a]
DR 1	6.9	4.9	6.3	Bw35,B27	—
DR 2	13.4	15.4	20.0	B7,Bw52	MS
DR 3	10.8	17.3	1.6	B8	JOD,CD
DR 4	9.6	4.9	23.5	B15	JOD,RA
DR 5	10.3	13.3	2.2	—	JRA
DRw6	2.2	5.3	4.6	Bw38	
DR7	12.5	9.8	0.5	A29,B13,B17 Cw6	CD
DRw8	2,7	5.6	6.5	—	JRA
DRw9	1.1	2.7	12.2	—	
DRw10	0.7	1.9	0.6	—	
Blank	29.8	19.0	22.0	—	

[a]MS = Multiple Sclerosis; JOD = Juvenile Onset Diabetes; CD = Celiac Disease; RA = Rheumatoid Arthritis; JRA = Juvenile Rheumatoid Arthritis.

subgroups included in the antigen DRw6 as it was defined during the Seventh Workshop. It is possible that the supertypic specificities MT1 and MT2 maybe helpful in dissecting these subtypes of DRw6. The definition of DRw6 is so difficult because of the wide range of cross-reactivity displayed by this antigen or group of antigens. The main cross-reactive pattern in DRw6/DR1/DR2. Another one is DR5/DRw6, and a third one is DR3/DRw6.

DR7 is a very well-defined antigen without any evidence for heterogeneity in spite of cellular data suggesting a splitting of Dw7. DR7 displays two major cross-reactive patterns: one with DR3 and a second one with DR4. There is a strong linkage disequilibrium between DR7 and several other HLA antigens: A29, B13, Cw6, and Bw57. DR7 is also involved in the disease association with celiac disease and dermatitis herpetiformis.

The antigens DRw8, 9, and 10 are relatively poorly defined new specificities. They have a fairly low frequency except for DRw9 in the Oriental population.

The gene frequencies, linkage disequlibria, and disease-associations of the HLA-DR antigens are summarized in Table 1.

C. Cross-Reactivity

One of the most intriguing observations in HLA serology is the extensive cross-reactivity found for alleles coded for by the same locus. In the HLA jargon, the term "cross-reactivity" is—somewhat sloppily—used to describe the phenomenon that an individual immunized with an antigenic difference of, per example, HLA-A1, will produce an antiserum which reacts not only with A1 but also with A10- and A11-positive cells. Also, pregnancy-sera frequently contain antibodies directed against three and more alleles of the same locus. It is the experience of every HLA-serologist that certain groups of antibody specificities (such as anti-HLA-A1, A10 and A11) occur together in a pregnancy serum more frequently than would be expected on the basis of the frequencies of the corresponding antigens in the population.[12] The typical HLA "cross-reactivity" has been found almost exclusively for alleles belonging to the same locus. This phenomenon is common to the alleles of the HLA-A,B and DR loci. Absorption and elution experiments[13] have demonstrated that most HLA-A,B antisera—even those that appear to be operationally monospecific—contain a number of different antibody populations, reacting with the antigenic determinants, which are shared by several "cross-reacting" alleles. Blocking studies using Fab-fragments[14] have shown that per example two *different* anti-B5 antibodies react with *different* antigenic sites and by capping-cocapping experiments, Kourilsky et al.[15] found these antigenic sites to be located on the *same* molecule.

From the fact that Svejgaard et al.,[16] have demonstrated an additive cytotoxicity by antibody populations reacting with different determinants, it can be understood why in HLA serology simple cross-absorption experiments can *not* be used to determine monospecificity. Per example, an antiserum may contain two antibody populations: one against a determinant shared by A1, A10, and A11, and one against a determinant unique to A1. This serum will react with A1 and A11 positive cells because the antibody populations have an additive cytotoxicity. Absorption with A1 positive cells will remove all activity, and absorption with A11 positive (A1 negative) cells will remove one of the two antibody populations and leave one behind which is at a sublytic concentration. Thus the phenomenon of "cross-reactivity" and "cross-absorption" may be simply explained by multiple shared antigenic determinants and the presence of the respective antibody populations in the sera. This concept is also supported by genetic considerations: HLA antigens are large polypeptides and it is, therefore, not surprising that this molecule should exhibit a number of distinct antigenic sites. If the alleles at one locus have evolved from each other by mutation, it is clear that there should be widespread sharing of determinants by the alleles. Of course this concept does *not* exclude the possibility of true cross-reactivity due to biochemical similarity of different determinants. Such chance similarity, however, would not be expected to show as strong a restriction to alleles as the serologic "cross-reactivity" (i.e., sharing of determinants) does.

There is yet another possibility which is intermediate to the above discussed concepts. It may be that a determinant with a given amino acid sequence has a slightly different expression on different molecules due to differences in the tertiary structure. This might explain the observation of different binding characteristics with monoclonal antibodies reacting with two cross-reacting antigens.[17]

Although the molecular and genetic concept of HLA cross-reactivity by widespread sharing of determinants has been developed for the HLA-A and B antigens, there is no reason why the same explanation should not be true for HLA-DR "cross-reactivity," where the serological observations (inclusion phenomena, cross-absorption, reaction patterns of monoclonal antibodies, etc.) are exactly the same as in HLA-A,B serology.

D. Supertypic (Public) Specificities

The phenomenon of supertypic specificities is closely connected with "cross-reactivity." As discussed above, one of the manifestations of "cross-reactivity" is the frequent joint occurrence of groups of antibody specificities in sera. This is due to the production of antibodies directed against antigenic determinants shared by several alleles. Such antisera have been grouped together in the clustering program (as described above) and were therefore considered by some to define new antigens. The comparison with the established DR antigens, however, shows very clear inclusion phenomena: MT1 includes the DR specificities DR1, DR2, and DRw6, the specificity MT2 includes DR3, DR5, DRw8, and DRw6; while MT3 includes DR4 and DR7 as well as possibly DRw9. Since MT1 and MT2 show a significant degree of overlap (both include DRw6), it is impossible to construct an allelic series from these three supertypic specificities. If, however, the cross-reactive families are arranged somewhat differently—and this can be done very easily by simply selecting the appropriate antisera—then one can construct a series which appears to be allelic: the specificity MB1 (including DR1, DR2, and DRw6), the specificity MB3 (including DR4 and DR5).

Since each of these supertypic groups, MB1, MB2, and MB3, are composed of two or more alleles of the DR series, it is not surprising that again the supertypic specificities would behave as alleles. It is quite obvious, that with a series of now 10 fairly well-established alleles of the DR locus, it would be possible to artificially create a very large number of different allelic series by combining different alleles of the DR series. (In fact there will be probably enough combination possibilities to supply a large number of famous HLA serologists with their own "new loci"). As the history of HLA-AB serology shows, such artificial loci tend to have a short life. There are some data by Tosi et al.,[18] which seem to indicate that a specificity called DC1, which is very similar in distribution to the MT1, MB1 specificity, may reside on a different molecule than that which carries DR1 or DR2. These data—if they are confirmed by other working groups—might indeed prove the existance of additional loci coding for B cell-specific antigens. From the information available about the mouse I region, it is quite clear that additional loci must exist.

There are some reports on B cell-specific antigens which do not fit into the HLA-DR series and which do not include any HLA-DR antigens. Park and Terasaki,[19] as well as Ahern et al.[20] described a series Te21 to Te24. These antigens are defined by either single sera or very few sera tested in duplicate. For the antigen Te24, a family is presented in which a recombination separates DR7 from Te24. This family, however, shows *two* recombinations of the same type in two different children, which is a fairly unlikely event. In this case it should also be considered whether it is possible that the antiserum defining Te24 may not be composed of several different antibody populations, one reacting weakly with DR7 and/or DR3-positive cells and another antibody population directed at antigens which are not coded for in the HLA region. In fact, Curtoni et al.[21] found that this serum contained a significant activity against little s of the MNSs system. Thus it may be concluded that the serological evidence for the existence of additional DR-like loci in man is still rather uncertain.

It should be stressed that the existence of supertypic specificities may be quite helpful in practical serology, just as in the HLA-B serology the presence or absence of Bw4 and/or Bw6 is of great help in the determination of subtypes of B locus antigens.

E. Definition of DR Antigens by Monoclonal Antibodies

It is quite obvious that this new technology will have a major impact on the definition of HLA-DR antigens. At the same time it should not be expected that all problems of HLA-DR serology could be solved quickly by the use of monoclonal antibodies.

One of the problems of the use of monoclonal antibodies in HLA serology is that many antibodies do not bind complement and are therefore not cytotoxic in the regular microlymphocytotoxicity technique. Such antibodies may nevertheless be quite valuable if a binding assay is used. It is, however, much preferable to have monoclonal antibodies which work in the regular microcytotoxicity assay, because it is possible to investigate the fine specificity of such antibodies on a large panel (N = 200). Non-cytotoxic antibodies are usually screened in a binding assay on a relatively small panel of cell-lines, which is rather unsatisfactory for determination of specificity. Some investigators tend to overlook the fact that human cell-lines, even if they are homozygous for some HLA antigens, are not at all comparable with congenic strains.

When using monoclonal antibodies, it must also be kept in mind that the expected repertoire of monoclonal antibodies detecting DR antigens reflects the view of the mouse immune system. In other words, there may be polymorphic determinants which the mouse (in most cases, Balb C) is unable to recognize because itself possesses a similar or identical determinant. It may also be possible that the mouse recognizes a determinant which is not relevant in the human allogeneic situation.

The experience with the production of anti-DR monoclonal antibodies shows that many antibodies are produced which react with a nonpolymorphic determinant on the DR molecule (so-called framework determinants). Such antibodies are not helpful in the dissection of the DR polymorphism; they are, however, very valuable reagents for molecular investigations for the detection of DR antigens on T cell blasts and for control reagents in order to assess the purity of a B cell preparation.

Since HLA antigens are made up of multiple antigenic determinants which are widely shared among the alleles of the same locus, one should expect that the majority of the monoclonal antibodies directed against HLA antigens should be broadly reactive and supertypic to the already defined antigens. This situation may change when it becomes possible to use human myeloma-lines and lymphocytes from immunized human beings for fusion, because it will then be possible to select the individual to be immunized according to this HLA-DR type. Per example, for the production of anti-DR3 antibodies, one would select an individual with the two most strongly cross-reacting antigens, namely DR6 and DR7.

At present there is not very much information available in the literature concerning monoclonal antibodies reacting with polymorphic DR determinants. Only one such antibody was included in the Eighth International Histocompatibility Workshop 1980.[22] This antibody was shown to include the specificities DR3, DR5, and DRw6 and was, therefore, very similar to the supertypic specificity MT2, which also includes DR3, DR5, and DRw6. Lysostrip experiments by Garotta et al.,[23] have demonstrated that this antibody strips the DR3, the DR5, and the DRw6 antigens from the cell surface. Therefore this antibody reacts with a determinant which is shared by these three specificities. Both the supertypic specificity and the results of the Lysostrip experiments are in complete agreement with what would have been predicted on the basis of the above discussed molecular concept of HLA cross-reactivity.

F. The Relationship Between HLA-DR and HLA-D

If one wants to compare the HLA-Dw antigens and the DR antigens, it is necessary to briefly review the differences in typing methods. While typing for DR antigens is a positive identification of a gene product on the cell surface by the action of a specific antibody, HLA-D typing is accomplished by mixed leukocytes culture of cells proven to be homozygous for the HLA-D locus antigen in family studies (stimulators) together with the cell to be typed (responders). If the responder does not recognize the HLA-D antigen of the homozygous typing cell, it is assumed that the responder possesses the respective HLA-D antigen. Thus the definition of antigens in HLA-D typing with homozygous typing cells is a negative identification; the presence of an antigen is determined by the absence of a reaction. Besides a number of technical problems connected with the HLA-D typing, there is also the problem that at least one additional locus can be separated by recombination from HLA-D, which codes for stimulation in mixed locyte culture and which can cause restimulation in the primed lymphocyte test.[24] Thus homozygous typing cells represent not only one HLA-D determinant, but also other determinants which may or may not be recognized by the responder cell. Therefore a "typing response" in HLA-D typing will generally require identity not only for the HLA-D locus but also for one or more other MLC-stimulating loci on the HLA-chromosome. From this it becomes clear that typing with homozygous typing cells is a rather restrictive and nonsensitive technique.

The third method to detect antigens of the HLA-D region is the primed lymphocyte typing technique. In this technique lymphocytes are primed with a selected antigenic difference, and later restimulated with the cells to be tested. A positive reaction in the primed lymphocyte typing technique is marked by an early stimulation response. In other words, the PLT test is a positive identification of the D region antigens, where the specificity of the reaction can be influenced by proper selection of stimulator and responder of the priming.[25]

Both in the Seventh[26] and in the Eighth International Histocompatibility Workshops,[27] a comparison was made between HLA-D typing based on homozygous typing cells and HLA-DR typing with antisera. Analyzed were haplotypes which have been derived from large families so that there was an equivocal definition of DR and HLA-D antigens. In both studies there is a very high correlation between HLA-D and DR, but no identity. The 2 × 2 table in both sets of data shows that the HLA-D pattern is included in the HLA-DR pattern. This finding suggests that HLA-D and DR antigens could be coded for by the same locus. If HLA-D and HLA-DR would be two closely linked but separable loci, one would have to expect approximately equal numbers of D-positive and DR-negative and D-negative, DR-positive discrepancies. The actual discrepancies found, however, are almost exclusively DR-positive, D-negative exceptions. As mentioned above, the DR typing test is much more sensitive than the HLA-D typing procedure with homozygous typing cells, and therefore it is not surprising that HLA-D should be included by HLA-DR. This is also in agreement with the observation that HLA-DR typing and typing by the PLT technique are much closer correlated than HLA-DR and HLA-D. There is no reason why stimulation in the primary mixed leukocyte culture should be elicited by different antigens than in the secondary mixed leukocyte culture. For this reason, one may conclude that most likely HLA-D typing and PLT typing are detecting the same gene products.

Another strong argument in favor of HLA-D and DR being on the same gene products is the observation that the cross-reactivity patterns in DR serology are virtually identical with those observed in PLT typing. It appears rather unlikely that two different loci should keep the same cross-reactive patterns even if they should have resulted from duplication. The very exceptional observation of haplotypes which carry different HLA-D and DR antigens, as per example DR3 and Dw4, might be explained by the assumption of an intragenic recombination. Thus the available evidence at this time (and we do not include the observation that HLA-DR antibodies block the mixed leukocyte culture as this could be also effected by steric hindrance), indicates that HLA-D and DR antigens could be features of the same gene product, measured with different techniques of different sensitivity.

IV. BIOLOGICAL SIGNIFICANCE OF HLA-DR ANTIGENS

Most of the information about the biological function of Ia antigens is derived from experimental work in mice, rats, or guinea pigs. Very little information exists about the functional characterization of HLA genes in the human species, mostly because of the lack of the appropriate experimental situation (congenic strains, etc.). There is, however, a very strong analogy between the MHCs of all manuals so that it appears permissible to use information gathered in several species, also in man. Following this line of reasoning, there are at least three groups of functional aspects connected with DR antigens:

1. Immune response genes
2. Restriction of cooperation of macrophages, T cells, and B cells in the immune response
3. T cell factors in the regulation of the immune response

Immune response genes—Historically, Ia antigens were discovered in experiments designed to define the gene products of IR genes. Although it is still not unequivocally proven, that the Ia antigens are identical with the IR gene products, the possibility of at least a partial identity remains quite likely. In man there is unfortunately no solid evidence for IR genes. This is not very surprising in view of the fact that also in inbred mice the immune response may be a product of several complementing genes in *cis*-position. In the outbred human species, this situation must be much more complex and it is therefore expected to be extremely difficult to show experimentally the presence of human IR genes. The very strong analogy between the mouse and men MHC, however, suggests strongly that human IR genes do exist on chromosome 6. Furthermore, the DR antigens as presently defined could at least partially be identical with the human IR gene product.

Restriction of cooperation in the immune response—Ia antigens have been recognized as the target of restriction phenomena in the cooperation between T and B cells and between macrophages and T cells in the humoral immune response.[28] In man, restriction by DR antigens has been shown in the macrophage requirement for antigen presentation in the proliferative response of mixed leukocyte culture.[29] From the data in the mouse, and the relatively limited experience in man, it may be concluded that HLA-DR antigens must be recognized in the process of cellular cooperation in the immune response.

T cell factors in the regulation of the immune response—In the mouse systems, it has been shown that upon the appropriate antigenic stimulus, T cells can produce factors which influence the immune response in a positive way (helper factors) or in a negative way (suppressor factors). These antigen-specific factors have been shown to contain Ia antigens.[30] Although similar experiments are still missing in man, it maybe expected that DR antigens are also part of the antigen-specific helper or supressor factors.

Although the biological functions of Ia antigens are presented here in three different experimental categories, it may well be that these different systems are measuring the same functions of the Ia, and in this case of the HLA-DR antigens. It has been speculated that Ia antigens might represent a part of the so far elusive T cell receptor which could be formed by the fusion of a VH gene product with the Ia antigens, whereby the antigen binding site is formed by the V gene product of immunoglobulin and the other part of the molecule, which is formed by the MHC gene product, could have the function of a constant part. The function of the receptor molecule (V gene products + MHC products) may be varied by the variable choice of different subclasses of MHC products such as I-A or I-C products.

V. CLINICAL IMPORTANCE OF HLA-DR ANTIGENS

A. Disease Susceptibility Genes

There are a high number of different diseases which show a very strong correlation with HLA-DR antigen in the random patient population.[31] It is generally assumed that the association with HLA antigens is caused by a disease susceptibility gene in very close proximity of the HLA marker loci. The very close linkage is connected with a high degree of linkage disequilibrium for the alleles of these loci. Among the DR antigen associated diseases there are the following major groups:

1. A group of organ-specific autoimmune diseases (Graves disease, autoaggressive chronic hepatitis, juvenile diabetes mellitus, myastenia gravis, etc.). All these diseases are highly associated with DR3 and Dw3.
2. Multiple sclerosis is associated with DR2 and Dw2.

3. Chronic rheumatoid arthritis of the adult is associated with DR4.
4. Certain forms of juvenile rheumatoid arthritis are associated with DR5.
5. Juvenile onset diabetes mellitus and celiac disease form a particular category in that these diseases are associated with two different DR antigens. In the case of juvenile diabetes it is DR3 and DR4, and in the case of celiac disease it is DR3 and DR7. The analysis of these double associations has shown that there is a significant excess among the patients both in juvenile onset diabetes as well as in celiac disease of those patients who carry both associated antigens.

These data are in favor of an interaction of different HLA linked disease susceptibility genes. From the available evidence, it may be concluded that there is so far no cogent evidence suggesting that the DR antigens themselves function as disease susceptibility genes. On the other hand, it is not impossible that the DR antigens are indeed by themselves involved in the pathogenesis. At any rate the disease susceptibility genes for these disorders must be located in very close proximity of HLA-D. It is not unlikely that these susceptibility genes might be related in function to the other genes of the HLA-D region.

B. Kidney Transplantation

In kidney transplantation, the correlation between HLA-A, B antigen matching and clinical success has been largely a disappointment to clinicians and immunogeneticists for the following reason. There was a strong hope at the time of the discovery of human HLA-DR antigens that these antigens might provide a relatively convenient and efficient matching system of limited polymorphism. The first sets of retrospective analysis of HLA-DR compatibility in relationship to the clinical success have shown remarkable success of HLA-DR matching. In a large international study performed in the framework of the Eighth International Histocompatibility Workshop, however, there was, if anything, only a very slight correlation between HLA-DR matching and clinical success.[32] From this disappointing experience it becomes clear that matching for DR antigens is not the single and simple solution of the problem of donor-recipient selection. Much rather, DR matching is only part of a complicated system of donor selection for clinical transplantation.

REFERENCES

1. **David, C. S., Shreffler, D. C., and Frelinger, J. A.,** New lymphocyte antigen system (Lna) controlled by the Ir region of the mouse H-2 complex, *Proc. Natl. Acad. Sci. U.S.A.*, 70, 2509, 1973a.
2. **Mempel, W., Grosse-Wilde, H., Albert, E., and Thierfelder, S.,** Atypical MLC reactions in HL-A typed related and unrelated pairs, *Transplant. Proc.*, 5, 401, 1973a.
3. **Harms, K., Granditsch, G., Rossipal, E., Ludwig, H., Polymenidis, Z., Scholz, S., Wank, R., and Albert, E. D.,** *Coeliac Disease*, Tu, J., Hekkergs, M., and Pena, A. S., Eds., H. E. Stenfert Kroese, Leiden, 1974, 215.
4. **Van Leeuwen, A., Schuit, H. R. E., and Van Rood, J. J.,** Typing for MLC (LD) I. The selection of non stimulator cells by MLC inhibition test using SD identical stimulator cells (MISIS) and fluorescence antibodies studies, *Transplant. Proc.*, 3, 1539, 1973.
5. **Van Rood, J. J., Van Leeuwen, A., Parlevliet, J., Terijtelen, A., and Keuning, J. J.,** LD typing by serology IV. Description of a new locus with three alleles, *Histocompatibility Testing*, Kissmeyer-Nielsen, F., Ed., Munksgaard, Copenhagen, 1975a, 629.
6. **Van Rood, J. J., Van Leeuwen, A., Keuning, J. J., and Bussé van Oud, Alblas, A.,** The serological recognition of the human MLC determinants using a modified cytotoxicity technique, *Tissue Antigens*, 5, 73, 1975.

7. *Bull. WHO.*, 56, 461, 1978.
8. **Van Rood, J. J., van Leeuwen, A., and Ploem, J. S.**, Simultaneous detection of two cell populations by two colour fluorescence and application to the recognition of B cell (Ia like) determinants, *Nature (London)*, 262, 795, 1976.
9. **Bodmer, J. G., Pickbourne, P., and Richards, S.**, Ia-Serology, in *Histocompatibility Testing 1977*, Munksgaard, Copenhagen, 1978, 35.
10. German regional report, in *Histocompatibility Testing 1977*, Munksgaard, Copenhagen, 1978, 449.
11. **Terasaki, P. I., Ed.**, *Histocompatibility Testing 1980*, P. I. Terasaki, Los Angeles, 1980.
12. **Albert, E. D., Mickey, M. R., and Terasaki, P. I.**, A new approach to crossreactivity in the HL-A system, *Int. Symp. Standardization of HL-A Reagents*, Copenhagen S. Karger, Basel 1973, 156.
13. **Mueller-Eckhardt, C., Heinrich, D., and Rothenberg, V.**, Frequency and complexity of crossreactive HL-A antibodies. Elution studies with platelets. Int. Symp. standardization HL-A reagents, Ser. Immunobiol. Stand., 18, 171, 1972.
14. **Legrand, L. and Dausset, J.**, Serological evidence of the existence of several antigenic determinants, (or factors) on the HL-A gene products, *Histocompatibility Testing 1972*, Dausset, J. and Colombani, J., Eds., Munksgaard, Copenhagen, 1973, 441.
15. **Kourilsky, F. M., Silvestre, D., Neauport-Sautes, C., Loosfelt, Y., and Dausset, J.**, Antibody-induced redistribution of HL-A antigens at the cell surface, *Eur. J. Immunol.*, 2, 249, 1972.
16. **Svejgaard, A.**, Synergistic action of HL-A isoantibodies, *Nature (London)*, 222, 94, 1969.
17. **Parham, P.**, Monoclonal antibodies against two separate alloantigenic sites of HLA-B40, *Immunogenetics*, 1981, in press.
18. **Tosi, R., Tanigaki, N., Centis, D., Ferrara, G. B., and Pressman, D. I.**, Immunological dissection of human Ia molecules, *J. Exp. Med*, 148, 1592, 1978.
19. **Park, M. S. and Terasaki, P. I.**, Second DR locus in *Histocompatibility Testing 1980*, P. I. Terasaki, Los Angeles, 1980, 578.
20. **Ahern, A. T., Park, M. S., Artruc, S. B., and Fuller, T. C.**, Te22 and Te24 in *Histocompatibility Testing 1980*, P. I. Terasaki, Los Angeles, 1980, 583.
21. **Curtoni, E. S., Borelli, I., Cornaglia, M., Olivetti, E., and Peyretti, F.**, Antibodies for other blood systems present in Workshop DR sera and interaction with anti-DR antibodies, in *Histocompatibility Testing 1980*, P. I. Terasaki, Los Angeles, 1980, 900.
22. **Ceppellini, R., Garotta, G., and Trucco, M. M.**, Evaluation of xenogeneic monoclonal antibodies as HLA-typing reagents, *Histocompatibility Testing 1980*, P. I. Terasaki, Los Angeles, 1980, 924.
23. **Garotta, G., Barbanti, M., Calabi, F., Neri, T. M., Trucco, M. M., and Ceppellini, R.**, 8W1247: a xenogeneic monoclonal antibody against supertypic specificity DRw3,5 and 6. *Histocompatibility Testing 1980*, P. I. Terasaki, Los Angeles, 1980, 864.
24. **Zier, K. S., Grosse-Wilde, H., Huber, Ch., Braunsteiner, H., and Albert, E. D.**, Restimulation in secondary MLC by a non D-locus determinant within the MHC, *Immunogenetics*, 6, 459, 1978.
25. **Lamm, L. U., Cullen, P., Edwards, J. H., van Leeuwen, A., Larsen, B., Cann, H., Thompson, J., Albert, E. D., Monk, K., Richards, S., and Bodmer, W. F.**, *Family Analysis Histocompatibility Testing 1977*, P. I. Terasaki, Los Angeles, 1977, 279.
26. **Zier, K. S., Scholz, S., and Albert, E. D.**, Crossreactivity in PLT: the specificity of primed lymphocytes (PL) depends upon the HLA haplotypes of the priming cell and responder transplant, 1, 723, 1979.
27. **De Marchi, M., Varetto, O., Savina, C., Borelli, I., Curtoni, E. S., and Carbonara, A. O.**, Relationships between HLA-D and DR, *Histocompatibility Testing 1980*, P. I. Terasaki, Los Angeles, 1980, 893.
28. **Katz, D. H., Graves, M., Dorf, M. E., Dimuzio, H., and Benacerraf, B.**, Cell interactions between histoincompatible T and B lymphocytes. VII. Cooperative responses between lymphocytes are controlled by genes in the I-region of the H-2 complex, *J. Exp. Med.*, 141, 263, 1975.
29. **Thorsby, E.**, MHC-restriction in man, *Eur. Surg. Res.*, 1981, in press.
30. **Taussig, M. J., Munro, A. J., Campbell, R., David, C. S., and Staines, N. A.**, Antigen specific T-cell factor in cell cooperation. Mapping within the I region of the H-2 complex and ability to cooperate across allogeneic barriers, *J. Exp. Med.*, 142, 694, 1975.
31. **Bertrams, J.**, HLA and disease association, *Behring Inst. Res. Comm.*, 62, 1978.
32. **Opelz, G. and Terasaki, P. I.**, International histocompatibility workshop study on renal transplantation, *Histocompatibility Testing 1980*, P. I. Terasaki, Los Angeles, 1980, 592.
33. **Baur, M. P. and Danilovs, J. A.**, Population Analysis of HLA-A,B,C, DR and other genetic markers, *Histocompatibility Testing 1980*, P. I. Terasaki, Los Angeles, 1980, 955.

Chapter 4

MONOCLONAL ANTIBODIES TO HLA-DR ANTIGENS

Jorge E. Kalil and Marc Fellous

TABLE OF CONTENTS

I. INTRODUCTION

Murine monoclonal antibody production is now a popular technique used by many groups working on cell surface antigens. Monoclonal antibodies provide precise reagents for the isolation, analysis, and physiological role studies of these structures. Production of antibodies in vitro by means of somatic cell hybrids was first achieved by Köhler and Milstein in 1975, and it has opened a new approach to immunogenetics.[1]

Murine myeloma cells are hybridized to splenocytes of a mouse which has been preimmunized against a chosen antigen. Selected hybrids produce antibodies in vitro against this antigen; myeloma cells provide a multiplying capacity in the culture and the "machinery" for the production and secretion of large quantities of antibody. Sensitized lymphocytes from the immunized mouse provide the specificity of this antibody.

Cloned hybrids will produce an unique antibody-type molecule directed to a specific epitope. Thus, a specific and normalized antibody can be obtained in limitless quantities because it is produced by an immortalized cell line. Each clone represents one of the antibodies making up complex and multispecific serum without fastidious absorptions.

Hybridomas obtained by xenoimmunization may reveal two kinds of antibodies; those recognizing an allospecific determinant and called polymorphic antibodies, and those which recognize an isotypic determinant, called monomorphic monoclonal antibodies.[2]

So far, using multiparous and polytransfused alloantisera, the study of HLA-DR has encountered problems that hopefully monoclonal antibodies will help to resolve.

The identification and characterization of immune sera recognizing HLA-DR antigens are detailed and time-consuming tasks. Multiparous sera must be absorbed on platelets so as to eliminate reactivity against HLA-A,B and C antigens and they must be screened on B lymphocyte-enriched cells.[3] Most of these sera are very cross-reactive and have a weak activity.

This makes more difficult the exchange of sera, the definition of new specificites, and also the study of cross-reactions. Do HLA-DR 1,2 and 6, for instance, form a real group of cross-reactions, or are these three antigens inherited in high linkage disequilibrium with another Ia-like molecule?[4] One major step in the dissection of the HLA complex has been the selective absorption of immune sera by well-genotyped HLA cells. However, absorption may bias considerably immunogenetic analysis. The main advantage of monoclonal antibodies is that they can replace such absorptions by the cloning of selected hybrids.

The production of polymorphic monoclonal antibodies recognizing described HLA-DR specificities would provide useful reagents for typing. They would not have either normal serum contaminants or normal extra-reactions, even without absorption. They could be produced on a large scale, and thus distributed throughout the world, obviously providing normalization for HLA-DR typing. They could be marked directly with an enzyme, a fluorescent substance, or a radioactive element, thus enabling other methods of tissue typing. As normal antisera are a mixture of antibodies with different affinity constants, and are directed to different epitopes, it is very difficult to approach the problem of the number of antigen molecules and the presentation of antigenic sites on the cell membrane. Further, it would be possible to approach the problem of synergy of antibodies directed to cell surface antigens in the fixing of complement to cause cell lysis.[5]

The number of loci coding for human Ia-like molecules is still unknown. The I region, murine equivalent of the human HLA-D/DR region, has 5 subregions defined by serological, biochemical, or functional analysis.[6] In the human system, two different populations of Ia-like antigens, probably coded by closely linked genes, have been described using classical antisera,[7–11] and two different loci controlling the proliferation in MLR have also been shown.[12,13] Highly specific monoclonal antibodies recognizing the product of just one subregion encoding for Ia-like antigens could provide an approach to these problems. They might also identify hybrid Ia-like molecules encoded in *cis* or *trans* position on the sixth pair of chromosomes. These hybrid molecules have already been demonstrated in the mouse system using serological and culture methods.[14-16]

Another question to be considered using monoclonal antibodies is whether molecules recognized serologically (HLA-DR), and those responsible for proliferation in an MLC (HLA-D), are one and the same.

The expression of Ia-like antigens on different populations or subpopulations of cells can be studied more easily using monoclonal antibodies. Strong and highly specific material may provide clarification of the HLA-DR antigens are expressed in a certain cell population. For instance, do nonactivated T lymphocytes express HLA-DR antigens in only very weak amounts?[17] The possibility of rapid and easy isolation of HLA-DR molecules using monoclonal antibodies may be very helpful in biochemical studies. Extensive studies in one- and two-dimensional gel electrophoresis and the construction of immunoabsorbants in order to purify HLA-DR antigens, thus allowing peptide map and amino acid sequence studies, will probably clarify the biochemical structure of these antigens.

II. PRODUCTION OF MONOCLONAL ANTIBODIES

A. Standard Technique

Production techniques are summarized in Figure 1.

A rat or mouse is immunized with whole human cells, membranes, or semipurified glycoprotein. Four days after the last injection, splenocytes are obtained and hybridized with a mouse or rat myeloma cell line maintained in culture in the exponential growth phase. The fusion agent used is polyethylene glycol (PEG) and the method of hybridization and culture based on that described by Köhler and Milstein and modified by other authors.[18-21] Different molecular weights and sources of PEG may vary in toxicity and fusion ability. Several HGPRT deficient myeloma cell lines are now available. These cells are resistent to 20 μg/ml of 8-Azaguanine, permitting hybrid selection in HAT (Hypoxanthine, Aminopterine, Thymidine) medium.[22] The mouse myeloma cell lines used are derivations of the BALB/c MOPC 21 myeloma. The first 8-Azaguanine variant described was P3/×63 − Ag8(×63) which secretes as MOPC 21, an IgG1 (k).[1] Some variants of ×63 have been described. The most frequent are: P3/NS1/1-Ag-4 (NS1) which synthesises the k-chain without its secretion, and SP_2O–Ag14 (SP_2O) which does not produce any Ig chain, nor does its clone FO.[21-24] The only rat myeloma line available is 210.RCY3.Ag.1.2.3 (Y3) from the Lou rat strain, which produces a k chain.[25]

Besides the production or not of Ig chains, these lines vary in their fusion ability. The proportion of splenocytes/myeloma cells may vary from 1/1 to 10/1.

Fusion is followed by distribution of the cells in flat-bottomed 96-well microtiter plates (200 μl per well). Two weeks after fusion and selection in HAT medium, only splenocyte × myeloma hybrids are growing in the wells. Supernates from well-developed colonies are then screened for antibody activity against HLA-DR antigens using any serological test. We used complement mediated microlymphocytotoxicity, complement fixation reaction, indirect immunofluorescence, a binding assay developed by iodinated *Staphylococcus aureus* Protein A, or an iodinated rabbit $F(ab')_2$ against mouse Ig.[26]

Table 1 shows an example of a screening of hybridomas produced by the fusion of

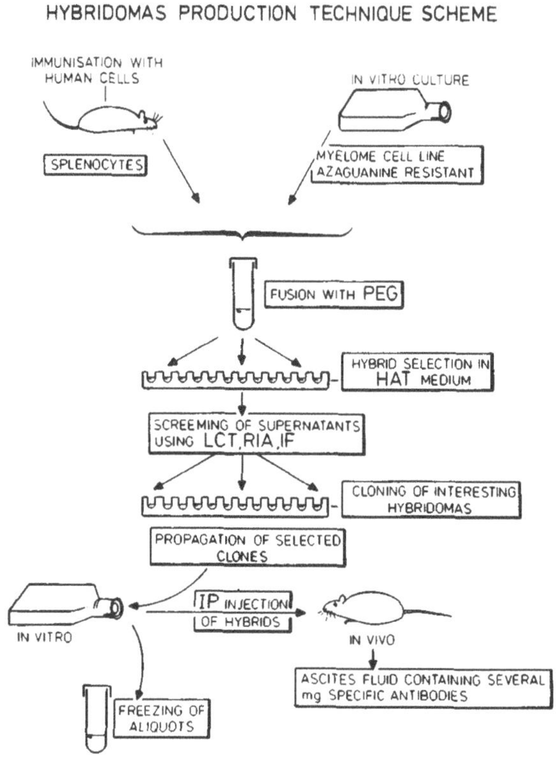

FIGURE 1.

the NS1 cell line with splenocytes from a mouse immunized against Mich (B lymphoid cell line HLA-A2, B15, DR5/5). Two tests were carried out: complement dependent microcytoxicity and a binding assay developed by iodinated Protein A, using a panel of well-phenotypes cell lines. From more than 120 independent colonies obtained, 75 supernatants showed a positive reaction against the immunizing cell Mich in at least one of the two techniques employed. The 14 colonies shown in Table 1 were chosen for a more detailed study because they had kept their activity 10 days after the first screening, or they seemed to recognize structures with a different expression depending on the cell type, when a small panel of cells was used (Mich-B cell line; T51 6.1.6-B cell line does not express HLA-DR antigens.[29] Daudi-B cell line expresses neither HLA-A,B antigen nor β_2m;[30] KE37-T cell line). This panel was selected as it gave an idea of the specificity of antibodies against MHC products (results not shown). Possibly interesting colonies were selected and cloned for further analysis. Cloning may be done by soft agar or limited dilution techniques.[27,28] Selected clones are propagated in cell culture and they can be injected intraperitoneally in syngeneic or nude mice to produce ascites containing some milligram of specific antibodies.

Unfortunately, both polymorphic monoclonal antibodies numbers 16 (anti-HLA-B15) and 31 (anti-HLA-DR5) lost their activity in culture. Numbers 2 and 18 were further characterized and showed that they recognized two epitopes of β_2m.[31]

B. Technical Problems Encountered in the Production and Characterization of Monoclonal Antibodies Against HLA-DR Antigens

The production of monoclonal antibodies against HLA-DR antigens raises many problems—these are either general production problems or those specific to the HLA-

Table 1
EXAMPLE OF SPECIFICITY STUDY ON SELECTED HYBRIDS OF MICH FUSION

				2	18	19	21	60	120	6	37	52	119	78	84	31	16
Supernatant number				2	18	19	21	60	120	6	37	52	119	78	84	31	16
binds complement-Protein				+	+	+	+	+	+	–	–	–	+	+	+	+	+
A-				+	+	+	+	+	–	+	+	+	–	+	–	–	+
Normal B cell lines	HLA-A	B	DR														
DS	1,3	35,27	1,1	+	+	+	+	+	+	+	+	+	+	+	+	–	–
COLIN	1,1	8,8	?,?	+	+	+	+	+	+	+	+	+	+	+	+	–	–
REMBI	2,2	15,15	2,2	+	+	+	+	+	+	+	+	+	+	+	+	–	+
LKT	1,1	8,8	3,3	+	+	+	+	+	+	+	+	+	+	+	+	–	–
PRIESS	2,2	15,15	4,4	+	+	+	+	+	+	+	+	+	+	+	+	–	+
MAITRE	24,28	51,45	5,5	+	+	+	+	+	+	+	+	+	+	+	+	+	–
MICH	2,2	15,15	5,5	+	+	+	+	+	+	+	+	+	+	+	+	+	+
ARNT	2,2	16,16	6,6	+	+	+	+	+	+	+	+	+	+	+	+	–	–
MANN	1,28	16,40	7,7	+	+	+	+	+	+	+	+	+	+	+	+	–	–
MADURA	2,2	40,40	8,8	+	+	+	+	+	+	+	+	+	+	+	+	–	–
T51	1,2	8,27	1,3	+	+	+	+	+	+	+	+	+	+	+	+	–	–
T51 6.1.6	1,2	8,27	–,–	+	+	+	–	–	–	–	–	–	+	+	+	–	–
T cell lines																	
JM	KE37	1301		+	+	+	–	–	–	–	–	–	–	–	–	–	–
MOLT 4	CEMT	JURKATT		+	+	+	–	–	–	–	–	–	–	–	–	–	–
Burkitt cell lines																	
DAUDI				–	–	–	+	+	+	+	+	+	+	+	+	–	–
CHE EBV –				+	+	+	+	+	+	+	+	+	+	+	+	–	–
CHE EBV +				+	+	+	+	+	+	+	+	+	+	–	–	–	–
Specificity hypothesis				anti β2m or HLA monomorphic?			monomorphic anti-DR						anti-B?	?		anti-DR5?	anti-B15

DR system. Some of the more pertinent points raised by our experiments are the following:

1. Immunization

The choice of animal for immunization can be important. Different rat or mouse strains might vary in their capacity to respond to HLA-DR antigens, as well as their ability to distinguish polymorphic determinants.

Most research groups have used whole cells as a source of antigen, and have obtained a majority of monomorphic monoclonal antibodies (see Table 2). Maybe the use of purified glycoprotein would provide monoclonal antibodies recognizing polymorphic determinants.[32] The use of intact membranes as antigen may give rise to cytotoxic antibodies, while soluble antigens would select noncomplement-fixing antibodies.[33]

Manipulation of the immune response using whole cells covered with selected monoclonal antibodies, which would hide most immunogeneic monomorphic determinants, may provide good results.[2]

2. Culture

The culture of hybridomas is much more delicate than that of lymphoid cell lines. Thus it is important to pay attention to the cellular density and culture medium. The use of feeder layers or conditioned medium for the first passage is advised. Thymocytes, peritoneal macrophages, or endothelial cell culture supernatant may be used for this purpose.[34]

Hybridoma productivity may be lost for two reasons: random loss of chromosomes by the hybrids, or the overdevelopment of nonsecreting clones. For these reasons, it is very important that interesting hybridomas be cloned very quickly.

3. Cloning

Cloning is very important in that it maintains antibody activity and also provides pure material for further specificity studies; a mixture of clones may mislead such studies. Cells must be cloned twice to assure monoclonality which can be confirmed by isoelectrofocalization of secreted Ig. The cloning of positive hybridomas replaces absorptions with a conventional serum. When a mouse is immunized against a whole human cell, it will produce antibodies directed against different determinants on the cell membrane (Figure 2). Thus in a conventional xenoantiserum, there is a mixture of different antibodies secreted by different cellular clones. After fusion and cloning, these antibody-secreting cells can be individualized and a very fine specificity can thus be defined.

4. Screening

Screening methods for hybridoma supernatants must be simple, sensitive, and rapid, enabling a very quick selection of positive colonies. The technique employed in screening must concord with the further utilization of the monoclonal antibody obtained; for example, microcytotoxicity or complement fixation reactions if the antibody is to be used for tissue typing, and an iodinated protein A binding assay if the antibody is to be used for biochemical studies—the property of fixing Protein A being very helpful for the characterization of antigens on the cell surface.[35] (It is perfectly viable that the same technique be used for screening and also for a later study. For example, cytotoxicity or an iodinated protein A binding assay can be used for tissue typing and later used for biochemical studies.)

It is essential to be aware of the major problems of sensitivity and specificity. False positive results may be obtained if the culture medium is not changed before screening;

Table 2
REPORTED ANTI-HLA-DR MONOCLONAL ANTIBODIES[a]

First author	Ref.	Name	Specificity DR	Ig class	Binds C′	Binds Prot. A	Immunizing antigen
Brodsky	41	DA2	Monomorph	IgG1	No	No	Membrane extract
		Genox 3.53	DR1,3,6	IgG1	No	No	Membrane glycoprotein
Charron	84	Ab1.35	Monomorph	IgG	Yes	Yes	B cell line
		Ab1.41	Monomorph	IgG	Yes	Yes	
		Ab2.06	Monomorph	IgG	Yes	Yes	
Trucco	2	S1.19	Monomorph	—[b]	No	—	B cell line
		S1.5	Monomorph	—	No	—	
		E15.4	DR3,5,6	—	Yes	—	
Quaranta	70	Q2/70	Monomorph	IgG1	No	—	B cell line
		Q5/6	Monomorph	—	—	—	
		Q5/13	Monomorph	IgG2a	Yes	Yes	
Reinherz	49	0K11	Monomorph	IgG2	Yes	—	Peripheral lymphocytes
Goyert	71	RC	Monomorph	—	—	—	B cell line
Lampson	67	L203	Monomorph	IgG	No	Yes	B cell line
		L227	Monomorph	IgG	No	Yes	
		L243	Monomorph	IgG	Yes	Yes	
Zola	36	FCM2	DR4		No	No	B cell line
Gross	68	D1.11	Monomorph	IgG2	Yes	—	Membrane fraction
		D1.12	Monomorph	IgG2	Yes	—	
		D4.22	Monomorph	IgG1	No	—	
Bono	44	L1.16	Monomorph	IgM	Yes	No	B cell line
Kalil	31	62.3.34	Monomorph	IgG2a	Yes	No	
		62.74	DR1	IgG2b	No	Yes	

[a] This list is not exhaustive. It includes the most used and well-characterized anti-DR monoclonal antibodies published to the present time (May 1981).
[b] Not shown in reference.

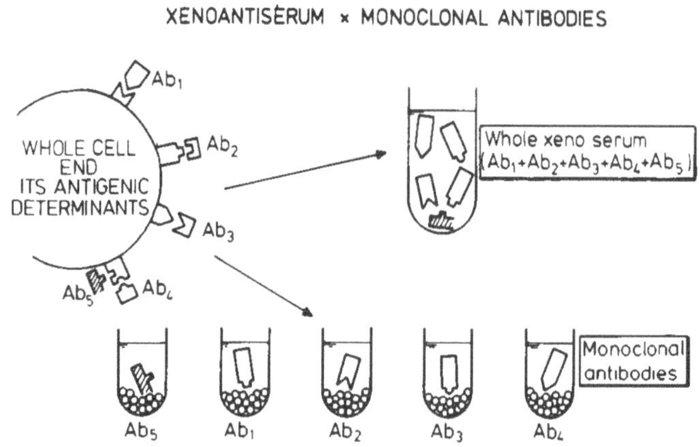

FIGURE 2. Xenoimmunization with a whole cell gives rise to a mixture of antibodies recognizing different determinants on the cell surface. Hybridization and cloning may individualize this population of antibodies by separating the productive B cell clones.

immunoglobulins secreted by unfused splenocytes may contaminate the supernatant during the first days of culture. False negative results may be obtained when the supernatant from an underdeveloped colony is taken for testing. A difference in the sensitivity of a given cell may give rise to false polymorphic supernatants.

III. MONOCLONAL ANTIBODIES AS TOOLS IN THE STUDY OF HLA-DR ANTIGENS

At least 10 groups have now published data on more than 20 anti-HLA-DR monoclonal antibodies (Table 2).

The majority of these hybridomas fix complement, and seem to react with the already described classical HLA-DR antigen. However, biochemical analysis has shown the complexity of these molecules in their reactions with these reagents.

The specificity of these monoclonal antibodies has been defined using cell distribution: activity on B cells and not on T cells. In the great majority of cases, this has been confirmed biochemically by immunoprecipitation, and analysis in PAGE.[36]

Several problems have been encountered in defining the specificity of these antibodies to Ia-like antigens.

1. The panel of lymphoblastoid cell lines carrying those HLA-DR specificities already defined is very limited, and thus analysis of a variety of B peripheral lymphocytes is necessary.
2. Complement-dependent assays and/or radioimmune binding assays have generally been used until now. However, there are classic pitfalls to these methods;
   a. Complement dependence is one of the major tests used in tissue typing. However, it raises the critical role of rabbit complement which is often contaminated with heterophile anti-human antibodies. It may in fact be counterproductive to use such precise and specific monoclonal antibodies with a complex reagent such as complement.
   b. Binding assay using anti-mouse Ig labeled with ^{125}I (or coupled with peroxydase or galactosidase) or purified Protein A ^{125}I, often leads to false, weakly positive activity, which may be related to the presence of Fc receptors on the B cells.

Table 3
PATTERN OF REACTIVITY OF CLONES 62.74, L1-1-6, AND 62.3.34 USING A BINDING ASSAY OR MICROCYTOTOXICITY ON HUMAN B AND T LYMPHOID CELL LINES

Lymphoid line	HLA-DR	Binding assay		Microcytotoxicity	
		62.74	control	L1-1-6	62.3.34
B cell lines					
DS	1/1	1572[a]	275	100[b]	100
Maja	1/1	2030	684	100	100
WOE	1/1	4435	769	100	100
JREE	1/1	1371	266	100	100
REMBI	2/2	308	261	100	100
SCTA	3/3	182	117	100	100
Goyen	3/3	605	377	100	100
Priess	4/4	1156	701	100	100
Maître	5/5	338	261	100	100
Mich	5/5	795	674	100	100
ARMT	6/6	371	284	100	100
MANN	7/7	1269	802	100	100
Madura	8/8	652	464	100	100
T51	1/3	4025	675	100	100
T51 6.1.6	—/—	735	585	< 5	< 5
T cell lines					
JM, KE37, Molt 4, CEMT, 1301	—/—	< 300	< 300	< 5	< 5

[a]cpm—positive results are underlined (> 2 times background)
[b]% cytotoxicity

Furthermore, individual cells may vary in the amount of antigen expressed on the surface of the cell. This increases the difficulties of comparison between different individuals tested. In an attempt to bypass this problem, McMichael et al.[37] recently proposed the use of an index ratio between two antibodies (polymorphic antibody anti-HLA/monomorphic anti β_2m) to better compare the reactivity of two cells.

A. Monomorphic Monoclonal Antibodies to HLA-DR Antigens

As we have already noted, most monoclonal antibodies described are of murine origin, and recognize isotypic antigenic determinants which are present on the HLA-DR molecule of all described allospecificities.

We have produced a monoclonal antibody (L1.1.6.) which recognizes a monomorphic determinant on HLA-DR antigens. It was produced by the hybridization of a BALB/c mouse immunized with LKT cells (HLA homozygous cell HLA-A1,B8,DR3) and NS1 myeloma cells. L1.1.6 is an IgM (k). Table 3 summarizes its pattern of reactivity in complement dependent microcytotoxicity against 15 lymphoid B and 5 T cell lines tested. It did not react with any of the T cell lines, but reacted with all the B cell lines tested, including T51, but not with its variant T51.6.1.6, which lacks HLA-DR antigens.[38]

Biochemical studies have confirmed these results. S^{35} Methionine biosynthetically

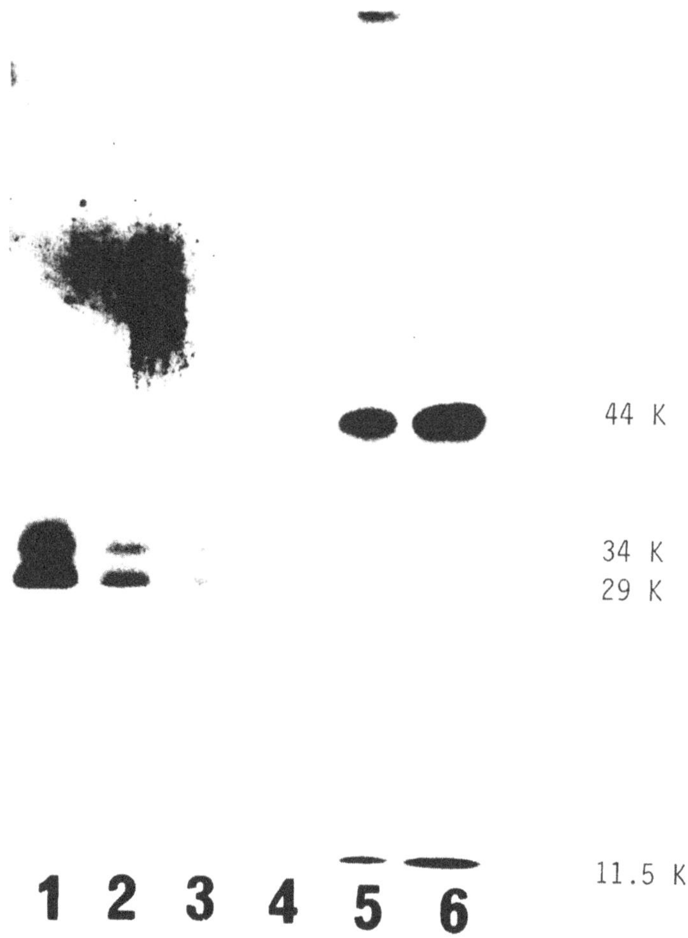

FIGURE 3. Immunoprecipitation with monoclonal antibodies. Immunoprecipitation on S^{35} Methionine internally labeled, NP40 solubilized, MAJA cells.

1. L1.1.6—anti-DR monomorph—mouse origin
2. 62.3.34—anti-DR monomorph—rat origin
3. 62.74—anti-DR1—rat origin
4. Normal mouse serum
5. M28—anti-β2m—mouse origin
6. M18—anti-β2m—mouse origin

labeled MAJA cells (HLA-DR1/DR1) and detergent solubilized membranes were used for immunoprecipitation with the L1.1.6. supernatant. Since L1.1.6. does not bind to protein A, an indirect method was used. A rabbit antiserum against mouse Ig was precipitated by formalin-fixed *Staphylococcus aureus* Cowan I. Immunoprecipitated material was analyzed on PAGE-SDS and Figure 3 shows the resulting pattern. Two bands of 29,000 and 34,000 daltons were immunoprecipitated; the β and α chain of DR antigens.[39]

Many research groups have quite easily obtained anti-HLA-DR in the mouse system, which might be expected when considering the important role played by Ia molecules in the immune response.[40]

B. Polymorphic Monoclonal Antibodies to HLA-DR

It seems much more difficult to isolate polymorphic monoclonal antibodies against HLA-DR molecules. Since monomorphic antibodies are of limited use in tissue typing, efforts are being made to produce antibodies which recognize allospecificities. Since it is clonal activation at the moment of immunization that determines the specificity of the antibody, different methods of immunization should be attempted in order to induce activation of murine B lymphocytes recognizing specifically polymorphic determinants on human HLA-DR antigens.

Purified membrane glycoprotein or highly purified Ia-like molecules could help in obtaining polymorphic monoclonal antibodies. Such an approach has been used for the production of antibodies against HLA-A2 and B7.[33] Brodsky et al. obtained a polymorphic monoclonal antibody, Genox 3.53, which recognized HLA-DR1,DR2 and DRw6 specificities—a classic group of cross-reactions—by immunizing C3H mice with purified membrane glycoprotein.[41] This antibody recognizes an antigenic determinant shared by these three allospecificities, or it can recognize DC1 antigens which are inherited in high linkage disequilibrium with these HLA-DR specificities.[8] Genox 3.32 is a polymorphic monoclonal antibody with one rare feature—it reacts with all HLA-DR specificities except DR7. This cross-reactive group has not yet been described using classical serological techniques.[33]

Trucco et al.[2] described a monoclonal antibody reacting with another classical group of cross-reactions in the serology of HLA-DR antigens: HLA-DR3, DR5, and DRw6.[2] FeM2 is another polymorphic monoclonal antibody described in a preliminary report by Zola et al.,[42] and it splits the HLA-DR4 specificity in two.[42]

It is interesting to note the difficulties encountered in obtaining cytotoxic, polymorphic anti-HLA-A,B, or DR monoclonal antibodies (see Table 2). Two explanations are possible: first, the subclasses of these antibodies do not bind the complement. Secondly, the low density of antigenic sites recognized by the antibody may be below the threshold necessary to trigger the complement cascade.[5] On the other hand, cytotoxic monomorphic monoclonal antibodies to Ia-like molecules would recognize products of distinct loci, allowing us to obtain the required density in order to promote complement lysis.

Howard et al.[43] described polymorphic monoclonal antibodies directed against MHC products in the rat, which were slightly or not at all cytotoxic. The mixture of two different monoclonal antibodies gave rise to a cytotoxic product provided that both antibodies were directed against two different antigenic determinants present on the same molecule.[43] Cellular lysis by one monoclonal antibody would then depend on the density of antigenic determinants recognized by this antibody.

In our laboratory, we have produced a hybridoma from the fusion of the splenocytes of a Lewis rat immunized with Raji cells (Burkitt lymphoma cell line) and mouse myeloma cells P3/X63/Ag8.[5] The supernatant of hybridoma 62 reacted specifically with B-enriched lymphocytes bearing the HLA-DR1 specificity (Table 4) when tested on a panel of 40 blood donors. After 5 days of culture, the 62 supernatant changed its pattern of reactivity and we proceeded to cell cloning. Two kinds of clones were seen; 62.74 (the majority) and 62.3.34. 62.74 produced a noncytotoxic rat IgG2b (k), binding *Staphylococcus aureus* protein A. 62.3.34 produced a cytotoxic rat IgG2a, but did not bind protein A. The pattern of reactivity of these two antibodies when tested on a panel of cell lines is shown in Table 3. 62.74 supernatant, when tested in a binding assay

Table 4
PATTERN OF REACTIVITY OF SUPERNATANT FROM HYBRIDOMA 62 USING A MICROLYMPHOCYTOTIXICITY TEST ON HUMAN PERIPHERAL B-ENRICHED LYMPHOCYTES

	HLA phenotypes		
A	B	DR	Reactivity of hybridoma 62[a]
1,3	17,35	1,7	2
2,23	14,15	1,4	3
11,26	15,35	1,X	3
28,X	35,40	1,6	2
33,44	2,12	1,X	3
2,30	12,18	1,X	3
29,20	37,12	1,3	3
2,30	12,18	1,X	3
29,26	37,12	1,X	3
1,28	2,44	7,1	3
2,4	5,7	2,5	—
1,25	12,18	2,X	—
1,2	8,7	3,7	—
24,X	18,40	3,5	—
11,23	18,	3,6	—
2,X	15,22	4,6	—
2,26	15,18	4,X	—
2,11	38,X	6,X	—
32,25	19,21	5,7	—
1,2	12,37	5,X	—
2,28	22,40	5,7	—
2,29	12,27	3,7	—
26,28	13,40	5,7	—
1,23	12,8	3,7	—
2,26	51,8	3,7	—
3,1	35,8	3,X	—
28,30	35,18	3,7	1
11,26	32,8	3,7	—
3,29	7,50	2,7	—
28,32	44,X	6,X	—
2,32	12,X	5,X	—
32,24	5,35	5,X	—
11,29	38,12	7,X	—
1,32	8,40	X,X	—
23,1	12,8	7,X	—
3,3	5,12	5,7	—
24,24	35,12	5,X	—
2,24	12,X	2,4	2
23,24	44,35	5,X	—
2,X	8,18	5,X	—

[a] —:<15% cytotoxicity.
1: 15–30%.
2: 30–50%.
3: 50–80%.

developed with iodinated protein A, reacted preferentially with cells and expressed HLA-DR1 specificity at more than 3 times the background. All other allospecificities bound to less than twice the background of radioactivity.

62.74 reacted strongly with T51 (HLA-DR1/3), but not with its variant T516.1.6 which did not express HLA-DR antigens.

On the other hand, 62.3.34 was cytotoxic when tested against all the 14 B cell lines tested which represented HLA-DR specificities, but was not cytotoxic with T cell lines. Like 62.74, 62.3.34 reacted with T51 but not with its DR negative variant.

The biochemical confirmation that these antibodies recognize well-defined Ia-like molecules is shown in the PAGE-SDS analysis (Figure 3). S^{35}-Methionine biosynthetically labeled Maja cells (DR1/DR1) were detergent solubilized and used for the immunoprecipitation. L1.1.6, 62.74, and 62.3.34 (columns 1,2,3) immunoprecipitated respectively HLA-DR molecules composed of 2 chains of 29,000 and 34,000 daltons. Columns 5 and 6 show the immunoprecipitation and the same cell lysate when 2 monoclonal antibodies against β2m were used.[31] Here also, 2 chains were precipitated: one of 45,000 daltons (the heavy chain of HLA-A,B molecules), and the second of 12,000 daltons (β2m).

One possible explanation for the loss in specific anti-HLA-DR1 lymphocytotoxicity activity in hybridoma 62 may be the loss of balance between the noncytotoxic anti-DR1 clone and the cytotoxic anti-DR monomorphic clone. Different assays using variable proportions between the monomorphic cytotoxic 62.3.34 and the polymorphic noncytotoxic 62.74 monoclonal antibodies have failed so far to produce a cytotoxic and specific anti-DR1 mixture.

C. Studies of the Expression of Ia-like Molecules in Different Tissues and Cell Populations Using Monoclonal Antibodies

HLA-DR antigens were first described on B lymphocytes.[45,46] Various cell populations have also been shown to express DR antigens.[47] Some of these studies have been confirmed and the expression detailed using monoclonal antibodies. Since monoclonal antibodies have no serum contaminants and are absolutely monospecific as reagents, they are helpful in analyzing the expression of HLA-DR on several cell types at different stages of differentiation.

1. Blood Cells

The expression of HLA-DR antigens on stimulated T lymphocytes, first described using alloantisera, was further confirmed by studies using a monomorphic monoclonal antibody to HLA-DR.[48,49] Pretreatment of T cells by anti-DR monoclonal antibody plus complement does not abolish further expression of Ia-like antigens after cell stimulation by a lectin or antigen. Thus, the expression of DR on activated T lymphocytes depends on the synthesis stimulation of a resting population of T lymphocytes expressing little or no DR antigens, rather than the expansion of a small subset of DR+ T cells.[50] On the other hand, human cytotoxic T cells do not express HLA-DR antigens. When MLC stimulated lymphocytes are treated with monoclonal anti-DR plus complement, cytotoxic T cells are enriched in the remaining population.[51]

Ia-like antigens are expressed on several precursors of blood cells.[52–54] The expression of HLA-DR on granulocyte-macrophage committed progenitor cells has recently been confirmed by monoclonal antibodies.[55] Treatment of bone marrow cells with monoclonal antibodies and rabbit complement inhibits granulocyte-macrophage colony formation in soft-agar culture. Unfortunately, this kind of study does not eliminate completely the possibility of the killing or auxiliary cells which thus hampers the formation of colonies.

Separation of peripheral blood lymphocytes into two populations has been employed in the study of the expression of this antigen on NK and K cells.[56] This study indicates that the great majority of cells bearing the natural killer (NK) and antibody dependent

Table 5
EXPRESSION OF HLA-DR ANTIGENS ON MELANOMA CELL LINES

			Monomorphic monoclonal antibody		
Cell line	Degree of pigmentation	62.3.34 αDR	L1.1.6 αDR	M28 αβ2m	BI232 αHLA-A,B
BEUR	++++	5	5	100[a]	100
DORI	—	100	100	100	100
BELO	++	50	50	100	100
VALD	—	90	90	100	100
IGR 37	+++	10	10	100	100

[a]Approximate percentage of staining.

cellular cytotoxicity (ADCC) activities are recovered in the DR negative population. These data conflict with other reports where conventional HLA-DR xenoantisera were used, showing evidence of the presence of Ia molecules on cells with NK or K activity.[57,58] Since these two cell populations are defined by a function, and they may represent a mixture of cell types, particularly the K cells, more detailed work must be done before a definitive conclusion can be drawn.

2. Expression of HLA-DR Antigens on Melanoma Cells

The presence of HLA-DR antigens as membrane constituents in the majority of melanoma cell lines has been identified by two independent groups. These cell lines failed to stimulate in MLC even though they seemed to express HLA-DR antigens in amounts comparable to lymphoid B cell lines.[59,60] Biochemical studies on the nature of Ia-like antigens on melanoma cell lines using monoclonal antibodies show that these antigens have a profile similar to those on B cells.[61]

We have studied some melanoma cell lines by indirect immunofluorescence using a panel of monoclonal antibodies. Results are shown in Table 5.

Anti-β2m (M18) and an anti-HLA-A,B,C, monomorphic antibody (B1.23.2) stained 100% of all cell lines tested. Three out of five cell lines were positive with both of our two monomorphic monoclonal antibodies against HLA-DR. In one case, half of the cells were positive; two of these cell lines were negative.

It is interesting to note that the more pigmented the cells, the less they were reactive with the anti-DR monoclonal antibodies, even though reactive with anti-HLA-A,B,C and anti-β2m antibodies. Thus, if one assumes that production of melanine is directly related to differentiation, then HLA-DR antigens may be expressed in earlier stages in the differentiation of melanocytes. These very preliminary results must be confirmed by more accurate studies on a much larger number of melanoma cell lines.

3. Expression on Different Tissues

Williams et al. have studied the expression of HLA-A,B,C and DR antigens on various human tissues using monoclonal antibodies.[62] By means of absorption experiments, they showed the presence of large amounts of HLA-DR antigens in the kidney—as much as 90% of the amount of DR antigens seen in the human spleen. Nevertheless, they failed to point out the type of cells expressing these antigens. No staining was observed when sections of fresh frozen kidney were incubated with a monomorphic monoclonal antibody against HLA-DR and then treated with a fluorescein labeled rabbit anti-mouse Ig.

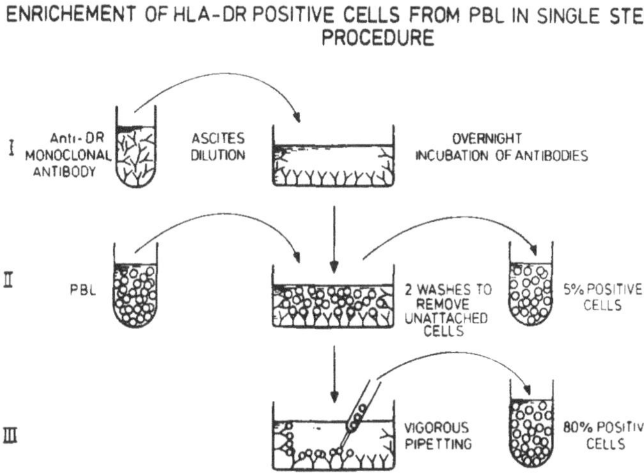

FIGURE 4. Petri dishes are incubated overnight with a dilution of an anti-DR monomorphic monoclonal antibody ascites (I). After washing to remove excess antibody, PBL are added and left to adhere (II). Nonattached cells are recuperated by washing with cold RPMI. DR+ cells are then recuperated by vigorous pipetting (III).

4. Estimation of the Number of HLA-DR Molecules

Trucco et al. used monoclonal antibodies to HLA structures to determine the number of antigenic sites present of the cell surface.[63]

The equilibrium constant of purified, quantitated, and radiolabeled monoclonal antibodies reacting with cell surface structures was determined graphically and then the number of reacting sites on the cell membrane was estimated. Using this technique, the expression of HLA-DR molecules was found to be 4.6×10^5 sites for a B cell line, and 0.8×10^5 sites for B cells in PBL.

5. Enrichment of DR+ Cells Using Monoclonal Antibodies

Monoclonal antibodies against HLA-DR antibodies enable the development of simple methods to enrich or deplete a cell population expressing Ia-like antigens.[64,65] This facilitates HLA-DR typing on DR enriched cells and permits the separation of cell subsets for functional studies. Stocker et al.[64] have developed a rosetting method using bovine erythrocytes coated with monoclonal antibodies against Ia-like antigens. Using this method, they passed from 20% to 75% of cells bearing surface immunoglobulin.

In our laboratory, we have developed another simple method for fractionating cells using monoclonal antibodies against Ia-like molecules (see Figure 4). Ascite fluids of anti-HLA-DR antibody L1.1.6 were diluted to 1/200 in RPMI supplemented with 10% FCS. Two ml of this dilution was put in Petri dishes (Falcon, 60 × 15 mm.) and left in contact overnight at 4°C. After three washes with RPMI, the plates were ready to use. A 2 ml suspension of peripheral lymphocytes (8.10^6/ml) was added to the plates and then incubated for 1 hr at 37°C. Cells in suspension were recuperated and after two washes with RPMI medium, attached cells were recuperated by vigorous pipetting. Two populations of cells were then tested in an indirect immunofluorescence assay for the presence of Ia-like antigens.

Using this method, it was possible to increase from 20% to 80% the number of cells expressing HLA-DR antigens.

D. Genetic Studies of Ia-like Antigens Using Monoclonal Antibodies

1. Immunological Dissection of the HLA-D/DR Region

Our knowledge of the genetic organization of the HLA-D region of the human chromosome 6 is very limited when compared with what has been described in the equivalent mouse I region.

Five different I subregions have been identified by functional, serological, and biochemical analysis called the I-A, I-B, I-J, I-E, and I-C subregions.[6] Two I subregions, I-A and I-E, have biochemically characterized products—Ia molecules. HLA-DR molecules are equivalent to the Ia molecules coded by I-E subregions when analyzed by peptide maps.[66] The human equivalent of I-A products is not yet known. Studies using selected multiparous antisera have suggested the presence of at least two series of Ia-like antigens in man.

Highly specific monoclonal antibodies recognizing gene products of different loci could help in providing an answer to these questions. Several research groups have been working in this field, using monoclonal antibodies.[67–71] Lampson and Levy described two monomorphic monoclonal antibodies recognizing two subsets of Ia-like molecules.[67] The complete immunoprecipitation of the Ia-like subset recognized by the first antibody did not interfere with the specific immunoprecipitation obtained by the second.

Shackleford et al., using the same antibodies in two-dimensional gel analysis, confirmed that they recognized different subsets of Ia-like molecules.[72] Nevertheless, they claimed that one subset was a precursor of the other, and that the two monoclonal antibodies recognized the same protein in different stages of maturation—they were thus not the products of two different genes. Gross et al., working with a radioimmunoassay using purified, iodinated Ia-like molecules, also dissected two subpopulations of Ia-like molecules.[68]

We have approached the question of different loci in the HLA-D region using two methods.[69,73] Firstly, we compared the reactivity of different monoclonal antibodies and a conventional multiparous anti-HLA-DR1 serum using co-capping experiments in indirect immunofluorescence on a homozygous DR1 cell (Table 6).

62.74, a polymorphic monoclonal antibody recognizing HLA-DR1 molecules, and L1.1.6, an anti-HLA-DR monomorphic monoclonal antibody, induced the complete capping of the molecules they recognized on the cell surface (columns 1 and 2). When 62.74 (anti-DR1) was used as the first reagent, followed by L1.1.6, there was not a complete co-capping. L1.1.6 recognizes other Ia-like molecules which are not recognized by anti-DR1 monoclonal antibody on a DR1/DR1 homozygous cell. The reverse experiment showed that L1.1.6 induced the complete capping of the molecules recognized by 62.74 (3 and 4).

These results were confirmed when L1.1.6 was compared with a conventional HLA-DR1 serum. L1.1.6 induced the capping of molecules recognized by the DR1 antiserum. The reverse experiment showed that L1.1.6 recognized molecules out of the cap formed by the DR1 antiserum.

These experiments suggest that at least two different antigen populations are present on the same cell surface of a homozygous DR1 cell. One population bears the DR1 specificity plus a monomorphic determinant recognized by L1.1.6, while the second population of antigens bears only the latter.

Co-capping experiments are very difficult to do using monoclonal antibodies, particularly with HLA-DR molecules. Several groups have encountered problems in capping DR molecules—there may be a problem of mobility on the membrane as DR molecules have two chains fixed in the membrane, while HLA-A and B antigens have only one. Another problem particularly concerns the use of monoclonal antibodies for this kind of study: complete saturation of membrane antigens is very difficult to achieve

Table 6
CO-CAPPING EXPERIMENTS COMPARING L 1.1.6, 62.74 AND A CONVENTIONAL DR1 ANTISERUM USING MAJA: B CELL LINE HLA-DR1/DR1

1st staining	2nd staining	1st staining[a] (%)	Capping[b] (%)	Co-capping[c] (%)	Nonco-capping[d] (%)
L 1.1.6	L 1.1.6	96	94	92	8
62.74	62.74	97	92	90	10
62.74	L 1.1.6	90	83	19	81
L 1.1.6	62.74	94	85	96	4
Anti DR1	L 1.1.6	85	79	40	60
L 1.1.6	Anti DR1	100	92	99	1

[a] $\frac{\text{number of stained cells}}{\text{total number of cells}} \times 100$

[b] $\frac{\text{number of capped cells (1st fluorochrome)}}{\text{number of stained cells}} \times 100$

[c] $\frac{\text{number of co-capped cells}}{\text{number of capped cells (1st fluorochrome)}} \times 100$

[d] 100 − co-capping

because there is only one antigenic site per molecule reacting with the antibody. At the moment of incubation at 37°C in order to promote the capping, the entropy of the system rises and the reaction equilibrium is deplaced to dissociation. Since X63 or NS1 are frequently used myeloma cell lines, hybrid antibody molecules may occur. These molecules have only one active site and the functional affinity resides only in the intrinsic affinity. All these problems make it difficult to perform interpretable co-capping experiments—all these parameters must be controlled beforehand.

Co-capping experiments have been confirmed by sequential precipitation. MAJA cells (DR1/DR1) were biosynthetically labeled with ^{35}S-methionine. Cell lysates were exhaustively immunoprecipitated with L1.1.6 and 62.74. The supernatant of these first precipitations was used for the second. The radioactivity precipitated is shown in Table 7. Results of the PAGE-SDS analysis are shown in Figure 5. Both antibodies precipitated specifically two chains of 34,000 and 29,000 daltons. After precipitation with L1.1.6, neither it nor 62.74 precipitated anything. On the other hand, when 62.74 was used first, L1.1.6 still precipitated Ia-like molecules. These results are in agreement with those of other groups working with monoclonal antibodies, and are not proof of a second locus of Ia-like antigens. These differences may simply be due to post-translational treatment. More accurate biochemical studies such as peptide maps, two-dimensional gel electrophoresis, amino acid sequencing, or more accurate familial studies to look for recombination between these putative two loci may help to solve this problem.

2. *Anti-HLA-DR Monoclonal Antibodies and Mixed Lymphocyte Culture*

Are HLA-D alleles determined by the mixed lymphocyte reaction (MLR) identical to HLA-DR alleles identified by serology? This problem can also be considered using monoclonal antibodies.

Table 7
SEQUENTIAL PRECIPITATION USING L 1.1.6 AND 62.74. MAJA (DR1/DR1) CELL LYSATES 5^{35} METHIONINE BIOSYNTHETICALLY LABELED

	First precipitation	Second precipitation	
		L 1.1.6	62.74
L 1.1.6	7500[a]	700	900
62.74	4800[a]	2700	800

[a]cpm.

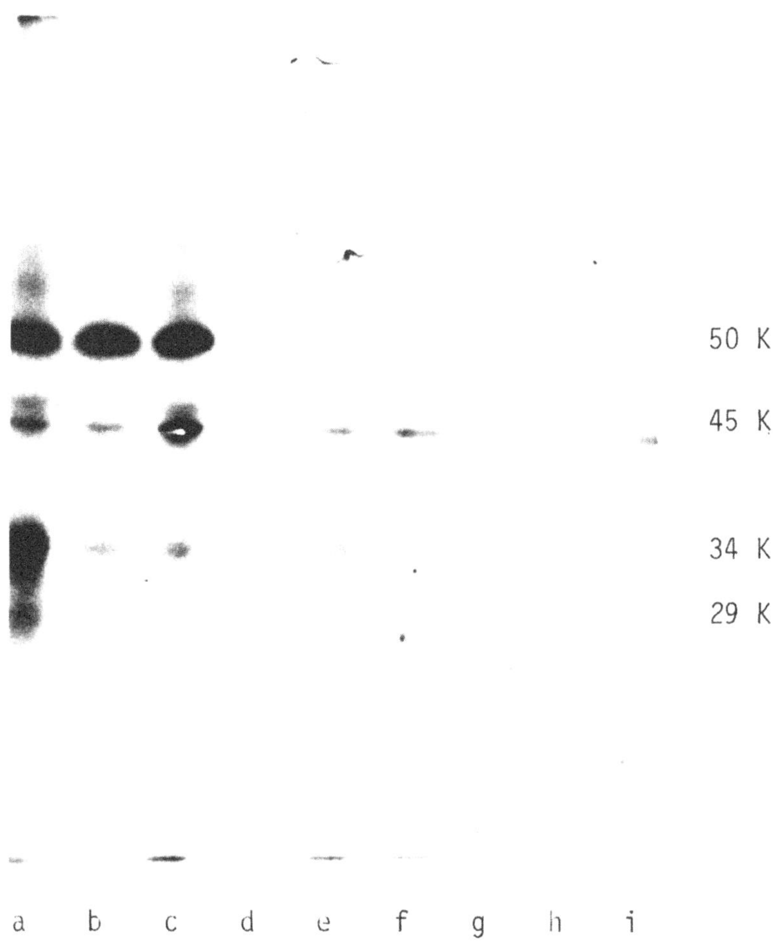

FIGURE 5. Sequential precipitation using L1.1.6 and 62.74. Autoradiography of PAGE-SDS from precipitates of L1.1.6 (monomorphic anti-DR—a), 62.74 (anti-DR1—b), and H (another monomorph anti-DR—c). Lanes d,e,f, show immunoprecipitation with L1.1.6 after pretreatment with itself, but it does precipitate DR molecules after treatment with 62.74 and H. Lanes g,h,i, show immunoprecipitation with 62.74 after pretreatment with the three anti-DR antibodies. 62.74 does not recognize anything further.

Blocking experiments could show the proliferative role of DR antigens in a primary or secondary MLR. When a conventional antiserum is used for this kind of study, one can imagine that hidden antibodies against HLA-D products may be responsible for the blockage. If monoclonal antibodies are used, this problem no longer exists. There are, however, two problems when an inhibition is present: the reaction of an anti-DR antibody may cause a steric hindrance of the HLA-D sites. Boyse et al.[74] showed that the fixation of an antiserum with certain antigens blocks the later reaction of a second antiserum directed to a neighboring antigen. Thus, Ly2 is adjacent to H-2D, Ly1 to H-2K, and H-2D to TL. Flaherty and Zimmerman confirmed these results excepting the last (H-2D and TL).[75] H-2D is not adjacent to TL when the cell is fixed with glutheraldehyde. They become adjacent after the reaction with an antiserum against H-2D or Tla. Another problem is the possibility of inhibition of proliferation by killing stimulating cells by ADCC (antibody-dependent-cellular-cytotoxicity).

Negative results do not exclude the role of HLA-DR antigens in proliferation in MLR. The monoclonal antibody may recognize an epitope which is not directly implicated in this reaction; low affinity of antibodies may permit dissociation, leaving molecules which are not hidden; and pinocytosis of reacting molecules may consume the inhibitor—all these phenomena may hamper MLR inhibition.

We have studied the inhibition of MLR I with 62.74, a monoclonal antibody which recognizes HLA-DR1 antigens. Using a Dw3 homozygous monoclonal cell as responder and 9 different stimulating cells, (5 expressing Dw1 and 4 non-Dw1) we obtained an average of 60% inhibition of MLR I against Dw1 cells, and 25% against non-Dw1 cells (data not shown).[44]

Indiveri et al.,[50] using PHA stimulated T lymphocytes as stimulating cells, showed that anti-HLA-DR monoclonal antibodies block the reaction. The same group also obtained a dose-dependent inhibition of allogeneic, but not xenogeneic, MLR with monoclonal antibodies.[76]

3. Cross-Reactions Between Species

Studies on the conservation of Ia-like antigenic sites in different species may aid us in our understanding of the evolution of these antigens. Four different combinations of cross-reactivities have so far been identified:

a. Polymorphic in man, not recognizing other species, such as Genox 3.53 and 62.74[41,77]
b. Polymorphic in man and in other species, e.g., polymorphic in rats and recognizes a cross-reactive group in man. MRC.OX.3 is such an example[94]
c. Monomorph in man and polymorphic in other species, as is the case for some anti-murine Ia monoclonal antibodies recognizing a polymorphism in the mouse and monomorphic in man[78]
d. Monomorphic in man and all primates tested, as is the case of DA2 and L1.1.6[41,77]

Another combination may be monomorph in man and not reacting with other primates. This has not been described in the HLA-DR system, but has been identified with β2m. M28 recognizes β2m in man and chimpanzee but not in other primates.[79]

E. Biochemical Studies of HLA-DR Using Monoclonal Antibodies

Monoclonal antibodies to MHC products have been used to purify membrane antigens in a soft and rapid procedure.

Since each monoclonal antibody differs in its affinity and susceptibility to eluting agents, many parameters may be attempted to find the best conditions. Garotta and

Trucco have described a mild method of elution of β2m; the temperature is increased from 0°C to 37°C.[80] Unfortunately, this method can only be used if the monoclonal antibody affinity constant shows drastic change between these temperatures. Herrmann and Mescher obtained a good yield of elution of H-2K molecules by changing to an ionic detergent at the moment of elution.[81] Acid elution seems to denature surface glycoproteins, thus while the mild methods used above cannot be employed, basic elution or ionic strength are the procedures of choice. Parham obtained a good conservation of protein structure of HLA-A,B,C using diethylamine 0.05*M* (pH 11.5), and Quaranta et al. obtained comparable results on HLA-DR using NaSCN 3M, with a lower yield.[82,83]

If the specific experimental purpose is to produce immunoabsorbents, then it is possible that high affinity antibodies are not the best to facilitate mild conditions of elution.

Another use of monoclonal antibodies for the biochemical characterization of HLA-DR molecules is study on two-dimensional gel electrophoresis. These studies have been performed by two independent groups.[84–86] Using this technique, it has been possible to correlate serological polymorphism with different patterns of gel migration and thus to identify the light (β) chain as responsible for allelic variations as has already been described by others using peptide maps.[87,88]

IV. DISCUSSION

The anti-DR monoclonal antibodies obtained so far have been particularly useful in advancing the state of knowledge of the genetic organization of the HLA-D/DR region, differential expression, and the biochemistry of Ia-like antigens.

The genetic organization of the HLA-D/DR region by its products using monoclonal antibodies has been shown to be highly complex. Several monoclonal antibodies have been described which react with Ia-like antigen subsets although recognizing the products of at least two different genes. There are two major problems with these studies. Firstly, to be sure that the nonreacting subset is really not recognized by the monoclonal antibody and that it is not just a problem of an insufficient dose, low affinity, or a diminution of affinity due to minimal changes produced in the antigenic preparation. The second problem is more difficult to solve. The possibility of the presence of immature proteins on the cell membrane increases the difficulty of an interpretation of the results.[89] The subsets of Ia-like molecules recognized by two different monoclonal antibodies may be produced by post-translational differences and not encoded by different genes. More detailed serological or biochemical studies must be carried out to clarify the genetic organization of this region. The use of selected alloantisera or polymorphic monoclonal antibodies in family studies to look for a recombinant would be the classical approach. Amino acid sequencing and peptide map analysis may show that Ia-like antigen subsets defined by monoclonal antibodies present, in fact, differences in the primary configuration of the protein. The real ordering of genes will probably be disclosed by gene cloning. Rapid progress has been made in this field with the cloning of HLA-A or B cDNA and the possibility of cell-free synthesis of HLA-DR antigens.[90,91] For the latter, monoclonal antibodies to separate α and β chains of DR antigens may be helpful.

The question of the possible identity of D and DR is not yet answered. The use of MLR blocking by monoclonal antibodies entails several problems as has already been discussed above. Fab fragments of high affinity monoclonal antibodies, directed to different epitopes of the HLA-DR molecule, should be used in large quantities to provide more convincing evidence that HLA-DR molecules are responsible or not for the proliferation in MLC.

Studies on the cross-reactions between species have shown that human monomorphic sites are also present in other primates and the monoclonal antibodies recognizing these sites may be polymorphic in the mouse. Polymorphic monoclonal antibodies in man either recognize polymorphic determinants in other species or they are not at all reactive.

These data may indicate preferential mutational target determinants in HLA-DR molecules. These determinants are not necessarily the same in distant species such as the mouse.

Another interesting point concerns antigenic sites and polymorphic monoclonal antibodies. Private determinants recognized by monoclonal antibodies are rare. 62.74 recognizes DR1 molecules but it still has a slight reaction with other described specificities. Other antibodies recognize well-defined antigens belonging to a cross-reactive group. This suggests that, in fact, there is an antigenic determinant shared by these antigens. Another explanation may be that these antibodies recognize just one antigen which is inherited in high linkage disequilibrium with the antigens described in the cross-reactive group. Another antibody described (Genox 3.32) indicates something not described by classical serology: the recognition of all DR specificities but DR7. This suggests that DR7 presents a clear mutation which distinguishes it from other specificities.

Monomorphic and polymorphic monoclonal antibodies to DR antigens have been produced in the mouse. The great majority of these monoclonal antibodies are monomorphic and this is perhaps the major problem. Other problems encountered in the production, characterization, and utilization of monoclonal antibodies concern their affinity and Fc properties. A large number of anti-DR monoclonal antibodies are IgG1 which does not bind complement or protein A: this problem does not occur with conventional serum where there is a mixture of Ig isotypes.

In order to bypass these difficulties, a rabbit antiserum to mouse immunoglobulin may be used, which enables, for instance, an indirect lymphocytotoxicity and an indirect immunoprecipitation.[41]

Antibody synergy experiments using monoclonal antibodies may clarify the conditions for obtaining cell lysis by complement. Reported data from Howard et al., and the possible synergy between 62.74 and 62.3.34 described here are two examples. The need for two immunoglobulins, closely fixed on the cell surface to trigger the complement cascade, may explain these results.[5] Thus, antigenic site density and fluidity on the membrane, as well as antibody affinity may control these events. Polymorphic anti-DR monoclonal antibodies would be less cytotoxic because HLA-DR antigenic sites are not dense or fluid. Anti-DR monomorphic monoclonal antibodies would recognize antigenic determinants of at least two different allelic series, attaining in this way the density threshold for complement lysis.

The expression of DR antigens on different cells in the different stages of differentiation has been carefully studied in an attempt to discern the biological role of these antigens. Hematopoietic stem cells express these molecules but only monocytes and B lymphocytes conserve them in the differentiated state. Peripheral T lymphocytes re-express in large amounts only after cell stimulation. Melanoma cell lines have also been shown to express DR antigens. These cells, of a completely different embryological origin, seem to express these antigens when undifferentiated. Several systems in positive and negative selections of DR bearing cells have been developed using monoclonal antibodies. Isolated populations may help in our understanding of why these antigens are expressed on these cell populations.

Anti-DR monoclonal antibodies fixing protein A have greatly eased the biochemical characterization of HLA-DR antigens. They have permitted extensive studies of these glycoproteins in one-dimensional and two-dimensional gel electrophoresis. The construction of immunoabsorbents permits the purification of HLA-DR molecules by rapid and simple means. This material can be used for structural sequencing studies. Purified material could be included in liposomes for physiological studies.

The highly sensitive screening techniques used may detect and conserve monoclonal antibodies with a very weak affinity. These monoclonal antibodies may cause problems when:

1. Peripheral lymphocytes are used in complement dependent cytotoxic tests to which lymphoblastoid cell lines are much more sensitive;
2. In indirect immunofluorescence and cocapping experiments where the second antibody layer may cause the dissociation of the first layer;
3. In immunoprecipitation and affinity chromatographic purification, where antigens are detergent solubilised and the antigen-antibody complex is subjected to a series of washes with different buffers to diminish contaminating substances.

On the other hand, high affinity antibodies may cause problems in elution in affinity chromatographic purification, thus necessitating the use of more rigorous procedures which may cause denaturation and a low yield.

The serologists' dream, born with hybridoma technology, has yet to be fully realized. We are not yet able to obtain large quantities of a very specific and discriminating reagent for cell typing.

Although much technical progress has been made since the first hybridomas, it is still random, and limited to obtaining polymorphic monoclonal antibodies for routine use. As the way in which the mouse "sees" surface antigens determines the monoclonal antibody specificities after fusion, many attempts have been made to specify the ideal antigenic site. The uses of purified protein, membrane extracts, whole cells covered with monoclonal antibodies, to hidden monomorphic determinants are examples of these efforts.

Apart from these assays with xeno-monoclonal antibodies, hopes greatly rose with the first good results using human monoclonal antibodies.[92,93] However, another cycle of experiments must be carried out with human cells to determine the optimum conditions for the predefinition of a desired specificity. With the advent of this new technology, it will perhaps be necessary to change current tissue typing techniques, permitting the use of antibodies that do not bind complement. Binding assay and ELISA (enzyme-linked-immunoabsorbant-assay) are practical and can be read automatically, but they still consume too many cells. Murine monoclonal antibodies have been very useful in demonstrating a lot of the cell surface markers present in cell subpopulations and not normally determining an immunological response in the same species. One example has been the rapid progress seen in the study of T lymphocyte differentiation membrane markers.[95] But the detailed dissection of allelic polymorphism or the distinction between two very similar polymorphic products encoded by closely linked genes, will probably be disclosed by allorecognition and human monoclonal antibodies.

Although the production of monoclonal antibodies of a predefined specificity has been undertaken since 1975, the great potential contribution of this technique to the study of cell surface antigens, and particularly of the HLA-DR type, has only just begun.

ACKNOWLEDGMENTS

We would like to thank Drs. François Hyafil and Denis Crevat for helpful discussion and aid; Dr. Donald Pious for providing the T51 6.1.6. cell line; Drs. Claude Mawas and Bernard Malissen for kindly providing B1.23.2 monoclonal antibodies; Dr. Nurit Saal for melanoma cells lines, and Mrs. V. Hawken for help in the preparation of the manuscript.

REFERENCES

1. **Köhler, G. and Milstein, C.**, Continuous cultures of fused cells secreting antibody of predefined specificity, *Nature (London)*, 256, 495, 1975.
2. **Trucco, M. M., Garotta, G., Stocker, J. W., and Ceppellini, R.**, Murine monoclonal antibodies against HLA structures, *Immunol. Rev.*, 47, 287, 1979.
3. **Ferrone, S., Allison, J. P., and Pellegrino, M. A.**, Human DR (Ia-like) antigens: biological and molecular profile, *Contemp. Top. Mol. Immunol.*, 7, 239, 1978.
4. **Bodmer, J. G.**, Ia antigens: definition of the HLA-DRw specificities, *Br. Med. Bull.*, 34, 233, 1978.
5. **Humphrey, J. H. and Dourmoshkin, R. R.**, Electron microscope studies of immune cell lysis, in *Complement—Ciba Foundation Symposium*, Wolstenholme, G. E. W. and Knight, J., Eds., J. & A. Churchill Ltd., London, 1965, 175.
6. **Uhr, J. W., Capra, D. J., Vitteta, E. S., and Cook, R. G.**, Organization of the immune response genes, *Science*, 206, 292, 1979.
7. **Van Rood, J. J., Van Leeuwen, A., Keuing, J. J., and Termiftelen, A.**, Evidence for two series of B-cell antigens in man and their comparison with HLA-D, *Scand. J. Immunol.*, 6, 373, 1977.
8. **Tosi, R., Tanigaki, N., Centis, D., Ferrara, D. B., and Pressman, D.**, Immunological dissection of human Ia molecules, *J. Exp. Med.*, 148, 1592, 1978.
9. **Duquesnoy, R. J., Marrari, M., and Annen, K.**, Identification of an HLA-DR associated system of B-cell alloantigens, *Transpl. Proc.*, 11, 1757, 1979.
10. **Katagira, M., Ikeda, H., Maruyama, N., Moriuchi, J., Wakisada, A., Kimura, S., Aizawa, M., and Itakura, K.**, Evidence for two cell alloantigen loci in the HLA-D region, *Immunogenetics*, 9, 335, 1979.
11. **Mann, D. L.**, Evidence for more than one locus controlling the expression of human B cell alloantigens (DR), *Fed. Proc., Fed. Am. Soc. Exp. Biol.*, 38, 1096, 1979.
12. **Mawas, C., Charmot, M., and Merceier, P.**, Split of HLA-D into two regions alpha and beta by recombination between HLA-D and GLO. Study in a family and primed lymphocyte typing for determinants coded in the beta region, *Tissue Antigens*, 15, 458, 1980.
13. **Shaw, S., Johnson, A. H., and Shearer, G. M.**, A new segregant series of human B cell alloantigens, *J. Exp. Med.*, 152, 565, 1980.
14. **Fathman, C. G., and Nabholz, M.**, In vitro secondary MLR. Detection of semi-allogeneic (F1) MLR stimulating determinants, *Eur. J. Immunol.*, 7, 370, 1977.
15. **Lafuse, W. P., McCormick, J. F., and David, C. S.**, Serological and biochemical identification of hybrid Ia antigens, *J. Exp. Med.*, 151, 709, 1980.
16. **Yokota, S., Lafuse, W. P., McCormick, J. F., and David, C. S.**, Detection of hybrid (combinatorial) Ia antigens using parent anti-F1 sera, *J. Immunol.*, 126, 371, 1981.
17. **Mann, D. L. and Sharrow, S. O.**, HLA-DRw alloantigens can be detected on peripheral blood T lymphocytes, *J. Immunol.*, 125, 1889, 1980.
18. **Köhler, G. and Milstein, C.**, Derivation of specific antibody-producing tissue culture and tumour lines by cell fusion, *Eur. J. Immunol.*, 6, 511, 1976.
19. **Gefter, M. L., Margulies, D. H., and Scharff, M. D.**, A simple method for polyethylene glycol-promoted hybridization of mouse myeloma cells, *Somatic Cell Genet.*, 3, 231, 1977.
20. **Kalil-Filho, J. and Fellous, M.**, Production d'anticorps monoclonaux à l'aide de la technique des hybridomes. Applications à l'immunogénétique, *Actualités Hématologiques (Masson, Paris)*, 14, 229, 1980.
21. **Fezekas de St. Groth, S. and Scheidegger, D.**, Production of monoclonal antibodies: strategy and tactics, *J. Immunol. Methods*, 35, 1, 1980.
22. **Littlefield, J. W.**, Selection of hybrids from matings of fibroblasts in vitro and their presumed recombinants, *Science*, 145, 709, 1964.
23. **Köhler, G., Howe, S. C., and Milstein, C.**, Fusion between immunoglobulin secreting and non-secreting myeloma cell lines, *Eur. J. Immunol.*, 6, 292, 1976.
24. **Shulman, M., Wilde, C. D., and Köhler, G.**, A better cell line for making hybridomas secreting specific antibodies, *Nature (London)*, 276, 269, 1978.

25. **Galfre, G., Milstein, C., and Wright, B.**, Rat × Rat myelomas and a monoclonal anti-Fd portion of mouse IgG, *Nature (London)*, 277, 131, 1979.
26. **Williams, A. F.**, Differentiation antigens of the lymphocyte cell surface, *Contemp. Top Mol. Immunol.*, 6, 83, 1977.
27. **Cotton, R. G. H., Secher, D. S., and Milstein, C.**, Somatic mutation and the origin of antibody diversity. Clonal variability of the immunoglobulin produced by MOPC 21 cells in culture, *Eur. J. Immunol.*, 3, 135, 1972.
28. **Lernhardt, W., Andersson, J., Coutinho, A., and Melchers, F.**, Cloning of murine transformed cell lines in suspension culture with efficiencies near 100%, *Exp. Cell Res.*, 111, 309, 1978.
29. **Gladstone, P. and Pious, D.**, Identification of a trans-acting function regulating HLA-DR expression in a DR-negative B cell variant, *Somatic Cell Genet.*, 6, 285, 1980.
30. **Fellous, M., Moltchelewicz, F., Kamoun, M., and Dausset, J.**, The use of a lymphoid cell line to define new B lymphocyte antigens, probably controlled by the MHC region, in *Histocompatibility Testing*, Kissmeyer-Nielson, F., Ed., Munksgaard, Copenhagen, 1975, 708.
31. **Kalil, J., Hyafil, F., Crevat, D., Bono, R., and Fellous, M.**, Monoclonal antibodies to MHC products, 4th Int. Cong. Immunol. Abstract Book, Preud'homme, J. L. and Hawken, V. A. L., Eds., S. F. I Paris, July 1980, Abstract No 8.1.17.
32. **Parham, P. and Bodmer, W. F.**, Monoclonal antibody to an human histocompatibility alloantigen, HLA-A2, *Nature (London)*, 276, 397, 1978.
33. **Brodsky, F. M., Parham, P., Barnstable, C. J., Crumpton, M. J., and Bodmer, W. F.**, Monoclonal antibodies for analysis of the HLA system, *Immunol. Rev.* 47, 3, 1979.
34. **Astaldi, G. C. B., Janssen, M. C., Lansdorp, P., Zeiglemaker, W. P., and Willems, C.**, in *Protides of Biological Fluids, 28th Colloquim*, Petter, H., Ed., Pergamon Press, London, 1980, 443.
35. **Kessler, S. W.**, Rapid isolation of antigens from cells with a Staphylococcal protein A antibody absorbent: parameters of the interaction of antibody—antigen complexes with protein A, *J. Immunol.*, 115, 1617, 1975.
36. **Laemmli, J. K.**, Cleavage of structural proteins during the assembly of the head of bacteriophage T4, *Nature (London)*, 227, 680, 1970.
37. **McMichael, A. J., Parham, P., Rust, N., and Brodsky, F.**, A monoclonal antibody that recognises antigenic determinant shared by HLA-A2 and B17, *Human Immunol.*, 1, 121, 1980.
38. **Kalil, J.**, Analyse des produits de la région HLA-D/DR à l'aide des anticorps monoclonaux. Thèse de l'Université Paris VII, 1980.
39. **Springer, R. A., Kaufmann, J. F., Siddoway, L. A., Giphart, M., Meo, D. P., Terhorst, C., and Strominger, J. L.**, Chemical and immunological characterisation of HLA-A linked B lymphocyte antigens, *Cold Spring Harbor Symp. Quant. Biol.*, 41, 387, 1977.
40. **Benacerraf, B. and Germain, R. N.**, The immune response genes of the major histocompatibility complex, *Immunol. Rev.*, 38, 70, 1978.
41. **Brodsky, F. M., Parham, P., and Bodmer, W. F.**, Monoclonal antibodies to HLA-DRw determinants, *Tissue Antigens*, 16, 30, 1980.
42. **Zola, H., Bradley, J., Macardle, P., McEvoy, R., and Thomas, M.**, A hybridoma antibody which reacts with human cells carrying the DRw4 tissue type. Evidence for heterogeneity of DRw4, *Transplantation*, 29, 72, 1980.
43. **Howard, J. C., Butcher, G. W., Galfre, G., Milstein, C., and Milstein, C. P.**, Monoclonal antibodies as tools to analyze the serological and genetic complexities of major transplantation antigens, *Immunol. Rev.*, 147, 139, 1979.
44. **Bono, R., Hyafil, F., Kalil, J., Koblar, V., Wiels, J., Wollman, E. E., Mawas, C., and Fellous, M.**, Monoclonal antibodies against HLA-DRw antigens, in *Transplantation and Clinical Immunology*, Vol. 11, Excerpta Medica, Amsterdam, 1979.
45. **Winchester, R. J., Fu, S. M., Wernet, P., Kunkel, H. G., Dupont, B., and Jersild, C.**, Recognition by pregnancy serums of non-HLA alloantigens selectively expressed on B lymphocytes, *J. Exp. Med.*, 141, 924, 1975.
46. **Humphreys, R. E., McCune, J. M., Chess, L., Malenka, D. J., Mann, D. L., Parham, P., Schlossman, S. F., and Strominger, J. L.**, Isolation and immunological characterization of a human B lymphocyte specific cell surface antigen, *J. Exp. Med.*, 144, 98, 1976.
47. **Schlossman, S. F., Chess, L., Humphreys, R. E., and Strominger, J. L.**, Distribution of Ia-like molecules on the surface of normal and leukemia human cells, *Proc. Natl. Acad. Sci. U.S.A.*, 73, 1288, 1976.
48. **Fu, S. M., Chiorazzi, N., Wang, C. Y., Mortazeri, G., Kunkel, H. G., Ko, H. S., and Gottlieb, A.**, Ia-bearing T lymphocytes in man. Their identification and role in the generation of allogeneic helper activity, *J. Exp. Med.*, 148, 1423, 1978.

49. **Reinherz, E. L., Kung, P. C., Pesando, J. M., Ritz, J., Goldstein, G., and Schlossman, S. F.,** Ia determinants on human T-cell subsets defined by monoclonal antibody, *J. Exp. Med.*, 150, 1472, 1980.
50. **Indiveri, F., Wilson, B. S., Russo, C., Quaranta, V., Pellegrino, M. A., and Ferrone, S.,** Ia-like antigens on human T lymphocytes: relationship to other surface markers, role in mixed lymphocyte reactions and structural profile, *J. Immunol.*, 125, 2673, 1980.
51. **Reinherz, E. L., Hussey, R. E., and Schlossman, S. F.,** Absence of expression of Ia antigen on human cytotoxic T-cells, *Immunogenetics*, 11, 421, 1980.
52. **Winchester, R. J.,** Aspects of surface membrane differentiation during maturation of human hematopoietic and tumor cells: the Ia system, in *Immunological Approaches to Embryonic Development and Differentiation, Part II*, M., Friedlander, Ed., Academic Press, New York, 1980, 115.
53. **Cline, M. J. and Billing, R.,** Antigens expressed by human B lymphocytes and myeloid stem cells, *J. Exp. Med.*, 146, 1143, 1977.
54. **Winchester, R. J., Ross, G. D., Jarowski, C. I., Wang, L. Y., Halper, J., and Broxmeyer, H. E.,** Expression of Ia-like antigen molecules on human granulocytes during early phase of differentiation, *Proc. Natl. Acad. Sci. U.S.A.*, 74, 4012, 1977.
55. **Fitchen, J. H., Ferrone, S., Quaranta, V., Molinaro, G. A., and Cline, M. J.,** Monoclonal antibodies to HLA-A,B and Ia-like antigens inhibit colony formation by human myeloid progenitor cells, *J. Immunol.*, 125, 2004, 1980.
56. **Ng, A. K., Indiveri, F., Pellegrino, M. A., Molinaro, G. A., Quaranta, V., and Ferrone, S.,** Natural cytotoxicity and antibody-dependent cellular cytotoxicity of human lymphocytes depleted of HLA-DR bearing cells with monoclonal HLA-DR antibodies, *J. Immunol.*, 124, 2336, 1980.
57. **Koide, Y., Takasugi, M., and Billing, R.,** B cell antigens on effectors for natural cell-mediated cytotoxicity, *Eur. J. Immunol.*, 7, 731, 1977.
58. **Chess, L., Evans, R., Humphreys, R. E., Strominger, J. L., and Schlossman, S. F.,** Inhibition of antibody-dependent cellular cytotoxicity and immunoglobulin synthesis by an antiserum prepared against a human B cell Ia-like molecule *J. Exp. Med.*, 144, 113, 1976.
59. **Winchester, R. J., Wang, C., Gibofsky, A., Kunkel, H. G., Lloyd, K. O., and Old, L. J.,** Expression of Ia-like antigens on cultured human melanoma cell lines, *Proc. Natl. Acad. Sci. U.S.A.*, 75, 6235, 1978.
60. **Wilson, B. S., Indiveri, F., Pellegrino, M. A., and Ferrone, S.,** DR (Ia-like) antigens on human melanoma cells. Serological detection and immunochemical characterization, *J. Exp. Med.*, 149, 658, 1979.
61. **Wilson, B. S., Indiveri, F., Molinaro, G. A., Quaranta, V., and Ferrone, S.,** Characterization of DR antigens on cultured melanoma cells by using monoclonal antibodies, *Transpl. Proc.*, 12, 125, 1979.
62. **Williams, K. A., Hart, D. N. J., Fabre, J. W., and Morris, P.,** Distribution and quantitation of HLA-A,B,C and DR (Ia) antigens on human kidney and other tissues, *Transplantation*, 29, 274, 1980.
63. **Trucco, M., de Petris, S., Garotta, G., and Ceppellini, R.,** Quantitative analysis of cell surface HLA structures by means of monoclonal antibodies, *Human Immunol.*, 3, 233, 1980.
64. **Stocker, J. W., Garotta, G., Hausmann, B., Trucco, M., and Ceppellini, R.,** Separation of human cells bearing HLA-DR antigens using a monoclonal antibody rosetting method, *Tissue Antigens*, 13, 212, 1979.
65. **Indiveri, F., Ng, A. K., Russo, C., Quaranta, V., Pellegrino, M. A., and Ferrone, S.,** Isolation of Ia-like antigen-bearing cells from human peripheral lymphocytes through the use of a monoclonal antibody to framework determinants of Ia-like antigens, *J. Immunol. Methods*, 39, 343, 1980.
66. **Silver, J., Walker, L. E., Reisfeld, R. A., Pellegrino, M. A., and Ferrone, S.,** Structural studies of murine I-E and human DR antigens, *Mol. Immunol.*, 16, 37, 1979.
67. **Lampson, L. A. and Levy, R.,** Two populations of Ia-like molecules on a human B cell line, *J. Immunol.*, 125, 293, 1980.
68. **Gross, N., Carrel, S., Tosi, R., Tanigaki, N., Carmagnola, A., and Acolla, R. S.,** Subsets of human Ia-like molecules defined by monoclonal antibodies, *Proc. 28th Colloq. Protides of Biological Fluids, Bruxelles*, Peters, H., Ed., Pergamon Press, Oxford, 1980, 595.
69. **Kalil, J., Hyafil, F., Bono, R., and Fellous, M.,** Immunological Analysis of HLA-D, DR Region Products Using Monoclonal Antibodies, paper presented to 8th Int. Histocompatibility Workshop, Los Angeles, January 1980.
70. **Quaranta, V., Tanigaki, N., and Ferrone, S.,** Distribution of antigenic determinants recognised by three monoclonal antibodies (Q2/70, Q5/6 and Q5/13) on human Ia-like alloantigens and on their subunits, *Immunogenetics*, 12, 175, 1981.
71. **Goyert, S. M., Hubert, J. J., Curry, R. A., and Silver, J.,** A novel molecule expressing HLA-DR antigenic determinants, *Human Immunol.*, 1, 161, 1980.

72. **Shackleford, D. A., Lampson, L., and Strominger, J. L.,** Analysis of HLA-DR antigens with monoclonal antibodies. Recognition of conformational differences in biosynthetic intermediate, *J. Immunol.*, 127, 1403, 1981.
73. **Kalil, J., Angles-Cano, E., Crevat, D., Sultan, Y., and Fellous, M.,** Production of monoclonal antibodies against HLA-DR antigens and the human plasma plasminogenactivator, *Pro. 28th Colloq. Protides of Biological Fluids, Bruxelles*, Pergamon Press, London, Peters, H., Ed., 1980, 587.
74. **Boyse, E. A., Old, L. J., and Stockert, E.,** An approach to the mapping of antigens on the cell surface, *Proc. Natl. Acad. Sci. U.S.A.*, 60, 886, 1968.
75. **Flaherty, L. and Zimmermann, D.,** Surface mapping of mouse thymocytes, *Proc. Natl. Acad. Sci. U.S.A.*, 76, 1990, 1979.
76. **Russo, C., Quaranta, V., Indiveri, F., Pellegrino, M. A., and Ferrone, S.,** Effect of polyclonal and monoclonal HLA-DR xenoantibodies on xenoantigeneic mixed lymphocyte reactions, *Immunogenetics*, 11, 413, 1980.
77. **Teillaud, J. L., Crevat, D., Chardon, P., Kalil, J., Goujet-Zalc, C., Mahouy, G., Vaiman, M., Fellous, M., and Pious, D.,** Monoclonal antibodies as a tool for phylogenetic studies of major histocompatibility antigens and β_2-microglobulin, *Immunogenetics*, in press.
78. **Pierres, M., Rebouah, J. P., Kovrilsky, F. M., Dosseto, M., Nercier, P., Mawas, C., and Mallissen, B.,** Cross-reactions between mouse Ia and human HLA-D/DR antigens analyzed with monoclonal alloantibodies, *J. Immunol.*, 126, 2424, 1981.
79. **Crevat, D., Kalil, J., Kadouche, J., and Fellous, M.,** Monoclonal antibodies to β2m, paper submitted.
80. **Garotta, G. and Trucco, M.,** HLA antigens temperature dependent elution from specific monoclonal antibody columns, in *Pro. 28th Colloq. Protides of Biological Fluids, Bruxelles*, Peters, H., Ed., Pergamon Press, Oxford, 1980, 463.
81. **Herrmann, S. H. and Mescher, M. F.,** Purification of the H-$2K^k$ molecule of the murine major histocompatibility complex, *J. Biol. Chem.*, 254, 8713, 1979.
82. **Parham, P.,** Purification of immunologically active HLA-A or B antigens by a series of monoclonal antibody columns, *J. Biol. Chem.*, 254, 8709, 1979.
83. **Quaranta, V., Walker, L. E., Pellegrino, M. A., and Ferrone, S.,** Purification of immunologically functional subsets of human Ia-like antigens by affinity chromatography on a monoclonal antibody immunoadsorbant, *J. Immunol.*, 125, 1421, 1980.
84. **Charron, D. J. and McDevitt, H. O.,** Analysis of HLA-D region-associated molecules with monoclonal antibody, *Proc. Natl. Acad. Sci. USA*, 76, 6567, 1979.
85. **Charron, D. J. and McDevitt, H. O.,** Characterization of HLA-D region antigens by two-dimensional gel electrophoresis. Molecular genotyping, *J. Exp. Med.*, 152, 18s, 1980.
86. **Shackleford, D. A. and Strominger, J. L.,** Demonstration of structural polymorphism among HLA-DR light chains by two-dimensional gel electrophoresis, *J. Exp. Med.*, 151, 144, 1980.
87. **Silver, J. and Ferrone, S.,** Structural polymorphism of human DR antigens, *Nature (London)*, 279, 436, 1979.
88. **Kaufman, J. F., Andersen, R. L., and Strominger, J. L.,** HLA-DR antigens have polymorphic light chains and invariant heavy chains as assessed by lysine-containing tryptic peptide analysis, *J. Exp. Med.*, 152 (suppl), 37s, 1980.
89. **Krangel, M. S., Orr, H. T., and Strominger, J. L.,** Assembly and maturation of HLA-A and HLA-B antigens in vivo, *Cell*, 18, 979, 1979.
90. **Ploegh, H. L., Orr, H. T., and Strominger, J. L.,** Molecular cloning of a human histocompatibility antigen cDNA fragment, *Proc. Natl. Acad. Sci., USA*, 77, 6081, 1980.
91. **Korman, A. J., Ploegh, H. L., Kaufman, J. F., Owen, M. J., and Strominger, J. L.,** Cell free synthesis and processing of the heavy and light chains of HLA-DR antigens, *J. Exp. Med.*, 152, 65s, 1980.
92. **Olsson, L. and Kaplan, H. S.,** Human-human hybridomas producing monoclonal antibodies of predefined antigenic specificity, *Proc. Natl. Acad. Sci. USA*, 77, 5429, 1980.
93. **Croce, C. M., Linnenbach, A., Hall, W., Steplewski, Z., and Koprowski, H.,** Production of human hybridomas secreting antibodies to measles virus, *Nature (London)*, 288, 488, 1980.
94. **McMaster, W. R. and Williams, A. F.,** Monoclonal antibodies to rat Ia-antigens: cross-reactions with mouse and human and use in purification of rat Ia glycoproteins, *Immunol. Rev.*, 47, 117, 1979.
95. **Reinertz, E. L. and Schlossman, S. F.,** The differentiation and function of human T lymphocytes, *Cell*, 19, 821, 1980.

Chapter 5

TISSUE DISTRIBUTION OF HUMAN Ia-LIKE ANTIGENS

Pier Giorgio Natali, Carlo Russo, Ah-Kau Ng, Patrizio Giacomini, Francesco Indiveri, Michele A. Pellegrino, and Soldano Ferrone

TABLE OF CONTENTS

I. INTRODUCTION

Analysis of the tissue distribution of human Ia-like antigens has been greatly facilitated by the development of monoclonal antibodies which have optimized immunofluorescence studies on frozen sections of tissues by eliminating background staining. During the last few years several studies have been published on this subject. Since this information contributes to our understanding of the role of these antigens in cell-cell interactions, in this chapter we will review the literature and our own data on the tissue distribution of human Ia-like antigens and we will compare it with that of Ia-like antigens in other animal species.

II. Ia-LIKE ANTIGENS ON HEMATOPOIETIC STEM CELLS

Expression of Ia-like antigens by stem cells (i.e., cells with an extensive capacity for self replication as well as the ability to differentiate into committed progenitor cells) and by progenitor cells committed to a specific line of hematopoietic differentiation has been mostly analyzed by incubating a bone marrow population with antibody and complement and then assaying for surviving stem cells by their ability to form hematopoietic colonies. An additional method used has involved staining bone marrow cells with antibodies and fluorescein-conjugated anti-Ig antibodies, separation of fluorescence positive and negative cells with a fluorescence activated cell sorter and testing the respective cell fractions for stem cells.

Moore et al.,[1] have suggested that human pluripotent stem cells lack Ia-like antigens on the basis of experiments performed with a long-term culture system in which proliferation of colony forming unit/granulocyte-macrophage is maintained for weeks to months and is believed to reflect feed-in from more primitive stem cells. According to these authors, treatment of bone marrow cells with complement and a xenoantiserum to Ia-like antigens did not reduce the number of colony forming unit/granulocyte-macrophage, when the surviving cells were cultured over an adherent layer derived from an initial culture of bone marrow cells. We have reached an opposite conclusion since incubation of bone marrow cells with monoclonal antibodies to Ia-like antigens and complement markedly inhibits the formation of mixed granulocytic and erythroid colonies from progenitor cells.[2] At present we cannot exclude that the conflicting results are caused by technical reasons, since in our experiments, the inhibitory activity of antibodies to Ia-like antigens may reflect their interaction with auxiliary cells, and the experimental conditions used by Moore et al.[1] do not exclude that the colony forming units/granulocyte-macrophage were generated from the adherent layer which had not been treated with antibodies and complement.

There is convincing evidence on the expression of Ia-like antigens by granulocyte-macrophage progenitor cells[1,3-7] and by eosinophil precursor cells.[8] Whether these antigens are expressed by erythroid progenitor cells is a matter of controversy. Winchester et al.[9] have reported that xenoantisera to Ia-like antigens and complement inhibit formation of colonies of hemoglobinized cells from committed erythroid progenitors, i.e., burst-forming (BFU-E) and colony-forming (CFU-E) unit/erythroid. Robinson et al.,[10] using monoclonal antibodies and cytofluorometry, detected Ia-like antigens only on BFU-E, suggesting a loss of Ia-like antigens when BFU-E differentiate into CFU-E. We have observed inhibition of BFU-E and CFU-E derived colony formation with xenoantisera to Ia-like antigens, but have not detected any inhibition with monoclonal antibodies.[11] If these results are not caused by a non-specific effect of xenoantisera, then the lack of activity of monoclonal antibodies may reflect either a very low density of Ia-like antigens on CFU-E and BFU-E, or an altered configuration which causes the lack of expression of the determinants recognized by the monoclonal antibodies used on the Ia-like antigen bearing molecules.

III. Ia-LIKE ANTIGENS ON CELLS OF THE MYELOID SERIES

The expression of Ia-like antigens on cells of the myeloid series is restricted to early stages of differentiation, as these antigens are found on the vast majority of myeloblasts and on a minor fraction of promyelocytes, but are not detectable on more mature granulocytes.[4,5] This information has prompted a reevaluation of the detection of Ia-like antigens on myeloid leukemic cells;[12–14] since the Ia-like antigen bearing cells in peripheral blood of patients with acute myeloid leukemia are mostly myeloblasts, it has been concluded that the expression of Ia-like antigens is an event related to hematopoietic differentiation[4,5] and not to leukemic transformation.[13] Furthermore, information about the distribution of Ia-like antigens on cells of the myeloid series has led to a reevaluation of the specificity of xenoantisera which had been raised with human leukemia cells and had been considered to recognize leukemia-specific antigens on the basis of the differential reactivity with human leukemic cells and mature granulocytes.[15] Analysis of the surface marker profile of myeloid cells at various stages of differentiation has shown that Ia-like antigens appear at an earlier stage than receptors for Fc portions of IgG and for complement components; IgG Fc receptors appear at the stage of promyelocyte and the number of positive cells increases through the myelocyte stage.[16,17] Receptors reacting with C3d become detectable at the stage of myelocyte and receptors reacting with C4d and C3b become detectable at the stage of metamyelocyte.[16,17] Therefore, the temporal relationship between appearance of Ia-like antigens and that of receptors for Fc portions of IgG and for complement components on cells of the myeloid series is different from that on cells of the lymphoid and monocyte lineages.[16,17]

IV. Ia-LIKE ANTIGENS ON MEGAKARYOCYTES

Up to 23% of megakaryocytes isolated from bone marrow are stained by rabbit antisera to human Ia-like antigens in indirect immunofluorescence.[18] It is not known whether the heterogeneity in the expression of Ia-like antigens reflects cells at differing stages of maturation or the development of separate megakaryocyte lineages. Differentiation of megakaryocytes to platelets is associated with loss of Ia-like antigens.

V. Ia-LIKE ANTIGENS ON LYMPHOID CELLS

Among cells of the lymphoid lineage, B cells express large amounts of Ia-like antigens;[19] their level is significantly increased on B lymphocytes following stimulation with PHA and transformation with Epstein-Barr virus. Indirect immunofluorescence staining of B lymphocytes from healthy donors and from patients with various types of lymphocytic leukemia with monoclonal antibodies to distinct determinants of Ia-like antigens has shown that B lymphocytes are heterogeneous in the expression of the antigenic determinants analyzed. Representative results are shown in Table 1. It is of interest that this heterogeneity is not restricted to the human system since we have also detected it by testing lymphocytes from chicken, guinea pigs, and rabbits with monoclonal antibodies to distinct determinants of human Ia-like antigens (Table 1). The latter series of experiments has also shown that the cross-reactivity of monoclonal antibodies to human Ia-like antigens with their counterparts in other animal species[20–23] extends even to sub-mammalian species, and that the ratio in the expression of distinct deter-

Table 1
REACTIVITY OF MONOCLONAL ANTIBODIES TO HUMAN Ia-LIKE ANTIGENS WITH PERIPHERAL BLOOD LYMPHOCYTES FROM VARIOUS ANIMAL SPECIES

Peripheral blood lymphocytes	# Positive/ # Tested	Monoclonal antibodies		
		Q5/13	Q5/6	Q2/80
Man	14/14	20(16–22)[a]	12(9–13)	8(5–10)
Monkey	4/4	48(42–52)	40(35–42)	50(46–54)
Rabbit	6/12	28(25–31)	4(2–5)	6(4–7)
Guinea pig	4/6	17(14–19)	15(14–18)	15(13–20)
Goat	1/1	25	38	6
Chicken	5/5	5(2–7)	47(44–52)	48(45–51)

[a]Percent of peripheral blood lymphocytes positive in indirect immunofluorescence.[22] Numbers indicate mean and range. Three hundred cells were counted for each determination.

minants identified by monoclonal antibodies varies among the animal species investigated. For instance, the monoclonal antibody Q2/80 reacts with a higher percentage of lymphocytes than the monoclonal antibody Q5/13 in chickens, while the reverse is true in man and the ratio changes from animal to animal in guinea pigs.

Analysis of the density of histocompatibility antigens on the B lymphoid cell line WI-L2 moving through the growth cycle has shown that the expression of Ia-like antigens is cell-cycle dependent being maximal at the G_2-M stage, while that of HLA-A,B antigens remains constant throughout the growth cycle.[24] These results, in conjunction with the differential effect of leukocyte interferon on the density of Ia-like antigens and HLA-A,B antigens,[25] suggest that these two types of histocompatibility antigens are controlled by different regulatory mechanisms.

There is general agreement that the majority of human T peripheral blood lymphocytes do not express Ia-like antigens, but there is controversy concerning the existence of a minor subpopulation of T cells bearing Ia-like antigens. Thus, Greaves et al.,[26] Yu et al.,[27] and Schuurman et al.,[28] using xenoantisera and an indirect immunofluorescence assay, have detected Ia-like antigens on about 4 to 10% of resting T lymphocytes. On the other hand, Evans et al.,[29] using a xenoantiserum and a ^{51}Cr cytotoxicity assay; Reinherz et al.,[30] using monoclonal antibodies and the fluorescence activated cell sorter; and ourselves[31] using xenoantisera, several monoclonal antibodies to distinct determinants of Ia-like antigens, and an indirect rosette microassay, have failed to detect Ia-like antigens on resting peripheral blood T lymphocytes. The conflicting results, if not caused by technical artifacts, may reflect differences in the methods of isolation of T lymphocytes, in the specificity of the reagents, and/or in the sensitivity of the assays. A significant number of Ia-like antigens bearing T lymphocytes can be detected in the peripheral blood of patients with immunodeficiency, infectious diseases, malignancies, immunopathologic diseases, and transplantation rejection.[27,32–36] In the patients investigated, no relationship has been found between the level of Ia-like antigens bearing T cells and clinical status.[36] In vitro studies have shown that the majority of T lymphocytes may acquire Ia-like antigens either following stimulation with mitogens,[29–31,37,38] soluble antigens,[38] and conditioned media,[32,39,40] or following culture with allogeneic B lymphocytes[29,30,37,41] and with autologous B lymphocytes infected with Epstein-Barr virus (Figure 1); the level of Ia-like antigens on PHA activated T

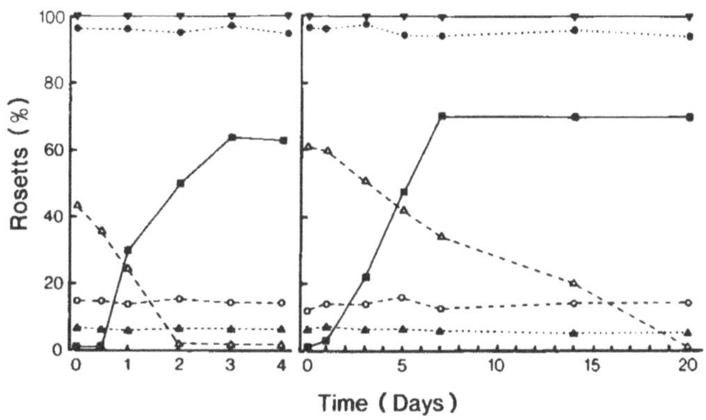

FIGURE 1. Kinetics of expression of Ia-like antigens and other surface markers on T cells stimulated with PHA (left panel) or isolated from cultures of peripheral lymphocytes infected with Epstein-Barr virus (right panel). T lymphocytes purified by nylon wool and cultured for various times with PHA were tested for membrane bound Ig (▲), Ia-like antigens (■), HLA-A,B antigens (▼), IgGFc receptors (○), IgMFc receptors (△), and sheep red blood cell receptors (●). The same tests were performed with purified T lymphocytes obtained at several interval times from cultures of peripheral lymphocytes infected with Epstein-Barr virus by depletion of B cells with red blood cells chemically coated with purified anti-human Ig.

cells is similar to that on autologous B cells.[42] According to Dobloug et al.,[43] even after overnight incubation in medium, up to 55% of T lymphocytes acquire reactivity with rabbit antisera to Ia-like antigens and with alloantisera to DR allospecificities, while under our experimental conditions, culture of T cells in medium does not effect their reactivity with monoclonal and polyclonal xenoantibodies to Ia-like antigens in serological assays.[31,44] We do not know whether the discrepancies reflect differences in the culture conditions, in the source of antibodies, and/or in the sensitivity of the assay systems. In this regard, it should be noted that there are significant variations in the percentage of Ia-like antigen bearing T lymphocytes found in PHA stimulated cultures of T lymphocytes isolated from a given donor on different occasions in spite of a similar degree of ^{3}H-thymidine uptake (Table 2).

The kinetics of appearance of Ia-like antigens on T cells stimulated with PHA or isolated from cultures of peripheral lymphocytes infected with Epstein-Barr virus is shown in Figure 1, which also contains data about the expression of other surface markers. In both systems the percentage of Ia-like antigens bearing T lymphocytes increases as a function of time in culture. In the cultures of lymphocytes with Epstein-Barr virus, Ia-like antigens become detectable on T lymphocytes following 5 days of incubation and the percentage of Ia-like antigens bearing T cells increases gradually until when it reaches a plateau on the 14th day of incubation. In cultures with PHA, in our own experience the increase in Ia-like antigen bearing T cells parallels the extent of ^{3}H-thymidine uptake and reaches a plateau after 72 hr of culture when the stimulation of DNA synthesis is maximal,[31] while according to Winchester and Kunkel,[45] it is delayed as compared to ^{3}H-thymidine uptake and reaches a plateau on the 7th day of culture. During the period of culture with PHA and with autologous B lymphocytes infected with Epstein-Barr virus, the expression of receptors for sheep red blood cells and the percentage of T cells expressing receptors for the Fc portion of IgG do not change, while the percentage of T cells expressing receptors for the Fc portion of IgM gradually

Table 2
LACK OF RELATIONSHIP BETWEEN THE PERCENTAGE OF T CELLS ACQUIRING Ia-LIKE ANTIGENS AND THEIR UPTAKE OF ^{3}H-THYMIDINE FOLLOWING STIMULATION WITH PHA

Exp.[a]	Addition of PHA	Ia-like antigen positive T cells[b] (%)	^{3}H-thymidine uptake (cpm)
1	+	82	49,075
	−	5	3,905
2	+	61	53,447
	−	6	4,127
3	+	45	45,135
	−	5	3,029

Note: Stimulation of purified T lymphocytes was measured after 6 days in culture by the uptake of ^{3}H-thymidine added 16 hr prior to harvesting.
[a]T lymphocytes were isolated from the donor F.I. on 3 occasions and were cultured in the presence and absence of PHA for 96 hr.
[b]Determined in an indirect rosette microassay[142] with monoclonal antibody Q5/13[141].

decreases and these receptors become undetectable following several days of culture. The newly expressed Ia-like antigens are not spatially associated with receptors for sheep red blood cells since T cells coated with anti Ia-like antigen antibodies maintain their ability to rosette with sheep red blood cells.[31] There are conflicting data about the spatial relationship between Ia-like antigens and receptors for Fc fragments of IgG and IgM: according to Samarut et al.,[46] Ia-like antigens are spatially close to Fc receptors since T lymphocytes coated with anti Ia-like antigen antibodies lose their ability to rosette with IgG coated erythrocytes. This relationship has not been confirmed by Dobloug et al.,[43] who, on the other hand, have found that IgG, but not Fab_2 fragments, of a rabbit antiserum to Ia-like antigens can inhibit the functional activity of receptors for IgM. The latter authors have interpreted these results to indicate that blocking occurs by steric hindrance caused by the Fc portion of IgG and that Ia-like antigens are spatially closer to Fcμ receptors on Tμ cells than to Fcγ receptors on adherent cells. Finally, using monoclonal and polyclonal xenoantibodies to Ia-like antigens and blocking assays, we have not found any relationship between Ia-like antigens and Fcμ as well as Fcγ receptors.[31] Furthermore, prior expression of Fcμ and Fcγ receptors is not a requirement for appearance of Ia-like antigens since isolated Tμ and Tγ subpopulations express Ia-like antigens following stimulation with PHA.[31]

Results by Ko et al.[38] suggested that expression of Ia-like antigens on activated T lymphocytes occurs because of a selective expansion of Ia-like antigen bearing T lymphocytes since treatment of T lymphocytes with a xenoantiserum to Ia-like antigens and rabbit complement significantly reduced the percentage of Ia-like antigens bearing T cells following stimulation with PHA. However, Reinherz et al.[30] and we[31] could not confirm this finding; we favor the possibility that the expression of Ia-like antigens on T cells cultured with PHA occurs by induction or increased synthesis.

The appearance of Ia-like antigens on T cells activated with mitogens is associated with their ability to stimulate allogeneic and autologous T lymphocytes in mixed lymphocyte reactions.[31,44] A role of Ia-like antigens in these reactions is indicated by (1) the relationship between the degree of stimulatory activity of PHA-activated T lymphocyte preparations and the percentage of Ia-like antigen bearing cells, and (2) the

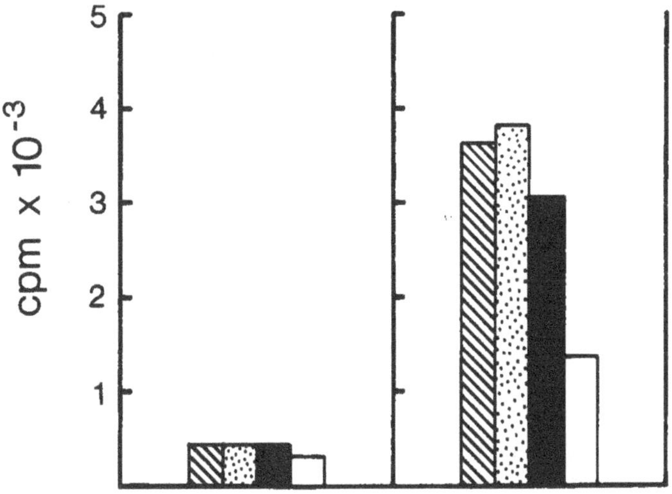

FIGURE 2. Effect of Ia-like antigen antibodies on unidirectional autologous mixed lymphocyte reaction. Purified T lympohcytes, from donor A.P., cultured for 72 hr in absence (left panels) and presence of PHA (right panels), were irradiated with 2000 rads and then used as stimulators in unidirectional MLRs with freshly purified autologous T lymphocytes. MLRs were performed in RPMI 1640 medium containing either fetal calf serum (▨), Ia-like antigen monoclonal antibody Q5/13 (■), HLA-A,B,C monoclonal antibody Q1/28 (▩), or rabbit Ia-like antigen antiserum 8802 (□). The rabbit Ia-like antigen antiserum was used at a final dilution of 1:50; normal rabbit serum at this dilution does not have an inhibitory effect on autologous MLR. Stimulation of responder cells was determined by their uptake of ^{3}H-thymidine.

ability of monoclonal antibodies to these antigens to specifically and significantly inhibit these reactions (Figure 2).

In view of the report[47] that antisera to the α chain of Ia-like antigens block autologous mixed lymphocyte reactions between T lymphocytes as responding cells and non-T cells as stimulatory cells, while antisera to the β chain do not, it is of interest that monoclonal antibodies recognizing determinants of the β chain of Ia-like antigens can inhibit stimulation of T lymphocytes by autologous PHA activated T lymphocytes.[44] These different results, if not caused by technical factors, suggest that Ia-like antigens play a role in both types of autologous mixed lymphocyte reactions, but distinct antigenic determinants induce proliferation of responding cells in the two systems.

Ia-like antigens on PHA activated T lymphocytes can be modulated with immunological and pharmacological approaches: when T cells are stimulated with PHA in the presence of monoclonal antibodies to Ia-like antigens, the number of T lymphocytes which acquire serologically detectable Ia-like antigens is significantly reduced, although their stimulatory activity in mixed lymphocyte reactions is comparable to that of T cells stimulated with PHA in the absence of antibodies (Table 3).[48] The effect of monoclonal antibodies to Ia-like antigens is specific and occurs very early in the process of PHA stimulation, since it is not seen if the monoclonal antibodies to Ia-like antigens are added 12 hr after the initiation of the cultures. It is likely that T lymphocytes activated with PHA in the presence of monoclonal antibodies to Ia-like antigens express these antigens at a level which is below the sensitivity of serological assays, but is still sufficient to stimulate responding cells in mixed lymphocyte reactions. The PHA induced expression of Ia-like antigens on T cells is more markedly inhibited by prednisone.[48]

Table 3
IMMUNOLOGICAL MODULATION OF Ia-LIKE ANTIGEN EXPRESSION AND STIMULATORY ACTIVITY IN AUTOLOGOUS MLR OF PHA ACTIVATED T LYMPHOCYTES[a]

T lymphocytes cultured in presence		Ia-like antigen positive T cells[b]	Autologous MLR[c] (cpm)		
PHA	MoAb	(%)	-	Q5/13	Q1/28
—	—	3	347	—	—
+	—	53	4132	1469	4828
+	Q5/13	5	3204	1341	4524
+	Q1/28	48	3497	1219	4024

[a]Purified T lymphocytes were stimulated for 72 hours with PHA in the presence of monoclonal antibodies Q5/13 to Ia-like antigens and Q1/28 to HLA-A,B antigens. T lymphocytes cultured without PHA were used as controls.
[b]Determined in an indirect rosette microassay[142] with monoclonal antibody Q5/13[141]
[c]Stimulation by autologous T lymphocytes was measured after 6 days in culture by the uptake of ^{3}H-thymidine added 16 hr prior to harvesting.

T lymphocytes from donors given a single oral dose of 12.5 mg of prednisone 6 hr prior to blood drawing display a marked reduction in PHA induced expression of Ia-like antigens as well as in stimulatory activity in autologous mixed lymphocyte reactions.[48] These parameters return to baseline values when T lymphocytes are isolated from blood drawn 24 hr following the administration of prednisone. Furthermore, the PHA induced expression of Ia-like antigens on T cells is influenced by the age of donors: T lymphocytes isolated from donors older than 70 years display a reduced expression of Ia-like antigens following in vitro stimulation with low amounts of PHA (Figure 3).

Structural analysis of Ia-like antigens synthesized by PHA stimulated T cells has shown that they have the apparent molecular weight of B cell derived Ia-like antigens;[30,31,49] the β chain of the T cell derived Ia-like antigens exhibits an increased mobility when electrophoresed under nonreducing conditions like that reported for the β chain of B cell-derived Ia-like antigens.[50,51] Furthermore, two-dimensional gel electrophoresis, peptide mapping analysis, and partial amino acid sequencing have indicated structural homology between Ia-like antigens synthesized by B lymphocytes and by autologous T cells activated with PHA.[49]

A large proportion of null cells express Ia-like antigens; these antigens are detectable on the majority of null cells which express receptors for the Fc portion of immunoglobulins and for the third component of complement.[52] The expression of Ia-like antigens by null cells is not required for displaying cytolytic activity in natural killer assay and in antibody dependent cell mediated cytotoxicity assays, since the cytolytic activity of null cells purified from lymphocytes through depletion of Ia-like antigen bearing cells is similar to that of null cells isolated through depletion of membrane-bound immunoglobulin-bearing cells.[53]

VI. Ia-LIKE ANTIGENS ON CELLS OF THE MONOCYTE-MACROPHAGE LINEAGE

While there is general agreement that in guinea pigs and mice Ia-like antigens are expressed only by a subpopulation of monocyte-macrophages,[54–58] conflicting data is

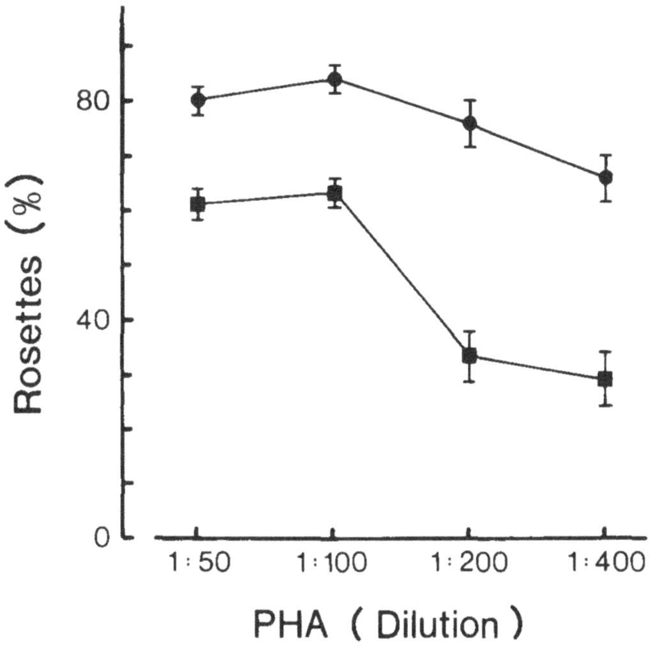

FIGURE 3. Effect of the age of donors on the expression of Ia-like antigens on T lymphocytes stimulated with PHA. T lymphocytes purified by nylon-wool procedure from 5 donors older than 70 years (■) and 5 donors younger than 40 years (●) were cultured for 72 hr with PHA at different concentrations. The expression of Ia-like antigens was determined by an indirect rosette assay using the monoclonal antibody Q5/13 to a framework determinant of Ia-like antigens. Points indicate mean (± range) of percent of rosettes obtained with cells from the 5 different donors.

available in the literature about the distribution of Ia-like antigens on human cells of the monocyte-macrophage lineage.

According to Winchester et al.[14,59] and McKinney et al.,[60] reaction of monocyte-macrophages with anti-DR alloantisera and xenoantisera to Ia-like antigens in indirect immunofluorescence stains all cells, although with significant differences in the degree of staining.[14] This finding was confirmed by several investigators[61–63] who tested peripheral blood monocytes with DR alloantisera in indirect immunofluorescence and cytotoxicity assays. On the other hand, Albrechtsen[64] showed that only up to 70% of peripheral blood monocyte-macrophages were lysed by anti-DR alloantisera and complement. We have reached a similar conclusion by analyzing monocyte-macrophages isolated from peripheral blood or from fluid of skin blisters induced with a 10% alcoholic extract of cantharidin (Table 4).[65] The level of Ia-like antigens on monocytes, as measured by the uptake of ^{125}I-SpA following incubation with an excess of monoclonal antibodies, is comparable to that on B lymphoid cells. Cytofluorometric analysis of monocytes stained with monoclonal antibodies to Ia-like antigens has shown that these antigens are mainly expressed on large monocytes[142] a subpopulation which displays significantly higher chemotactic activity than small monocytes.[66] Therefore, Ia-like antigens may represent a useful marker which, in combination with the available monoclonal antibodies and with cell sorter technology, can be used to fractionate monocyte subpopulations with distinct morphological and functional properties. Peripheral monocytes cultured in vitro continue to express Ia-like antigens;[65] this finding differs

Table 4
Ia-LIKE ANTIGENS ON HUMAN MONOCYTE-MACROPHAGES ISOLATED FROM PERIPHERAL BLOOD AND FROM EXUDATE OF SKIN BLISTER INDUCED WITH CANTHARIDIN

	Rosette[a] with monoclonal antibodies[b]		
Cells	Anti-Ia-like Q5/13	Anti-β_2-μ NAMB-1	Anti-MAA 376.96S
Donor Y.Z.			
Peripheral blood monocyte-macrophages	65	85	2
Blister exudate monocyte-macrophages	45	95	1
B lymphoid cell			
Raji	90	95	3
Myeloid cell			
K562	3	7	3

[a] Twenty μl of target cells (1×10^5) were mixed with equal volume of 0.5% suspension of red blood cells coated with either purified monoclonal antibody. The mixture was then spun at 500 g for 5 min. After resuspending the pellets, a minimum of 200 target cells were counted; target cells were counted as rosette positive if 3 or more SRBC adhere to them.
[b] Testing with the anti-human β_2-microglobulin monoclonal antibody NAMB-1 was used as positive control; testing with monoclonal antibody 376.96S to melanoma associated antigens was used as negative control.

from a recent observation in mice that biosynthesis and expression of I-A antigens by peritoneal macrophages terminate during the first day in culture.[67] It is not known whether these different results reflect differences in the biosynthesis of gene products of various subregions of the I region of the major histocompatibility complex or changes in regulatory mechanisms of Ia-like antigens occurring through phylogenetic evolution. Broxmeyer[68] has reported that the Ia-like antigen bearing monocyte subpopulation is sensitive to the inhibitory activity of lactoferrin on the production of colony stimulatory activity for granulocytes and macrophages. Monocytes coated with antibodies to Ia-like antigens become insensitive to the inhibitory activity of lactoferrin and Broxmeyer has interpreted these results as indicating that Ia-like antigens either act as receptors for lactoferrin, are spatially close to receptors for lactoferrin, or interact with these receptors following coating with antibodies.

In agreement with data in rodents,[69] Ia-like antigens synthesized by human monocyte-macrophages have a structure similar to that of B lymphoid cell derived Ia-like antigens (Figure 4) and play a role in the interaction between T lymphocytes and macrophages.[70] Lysis of Ia-like antigens bearing macrophages eliminates their ability to present antigens to T lymphocytes; macrophages must share at least one DR allospecificity with T lymphocytes. Antisera to the DR allospecificity shared between macrophages and T cells, as well as monoclonal antibodies to public determinants of Ia-like antigens, can inhibit the generation of an immune response.[65,70]

The expression of Ia-like antigens by monocyte-macrophages is associated with their ability to stimulate autologous and allogeneic lymphocytes in mixed lymphocyte reactions.[69,71,72] As shown in Figure 5, these proliferative responses are blocked by monoclonal antibodies to Ia-like antigens, but are not affected by antibodies to HLA-A,B antigens.[63,65]

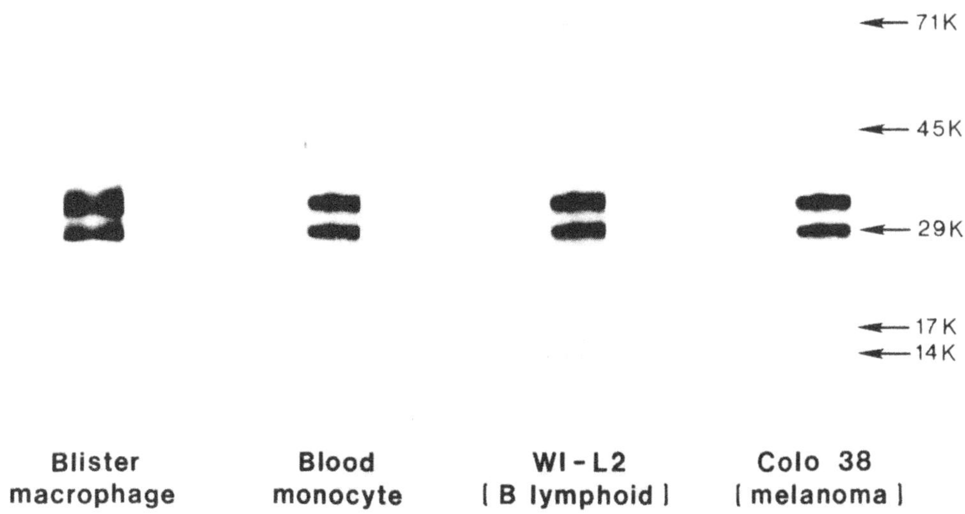

FIGURE 4. Molecular profile of Ia-like and HLA-A,B antigens synthesized by human peripheral blood monocyte-macrophages, skin blister monocyte-macrophages, B lymphoblastoid cells WI-L2, and melanoma cells Colo38. NP-40 extracts of cells labeled with ^{3}H-glucosamine were immunoprecipitated with the anti Ia-like antigen monoclonal antibody Q5/13. The labeled antigens were eluted with sodium dodecyl sulfate and 2-mercaptoethanol at 100°C and then electrophoresed on slab polyacrylamide gel as described.[20] The antigen molecules in the slab gels were detected by fluorography and their mobilities compared with those of standard proteins.

VII. Ia-LIKE ANTIGENS IN TISSUES OF NONLYMPHOID ORIGIN

Testing with a limited number of tissues initially suggested that Ia-like antigens have a restricted tissue distribution and are mainly expressed on cells associated with immune functions.[19] However, testing of lysates from various organs for their ability to bind conventional allo- and xenoantisera and monoclonal antibodies to human Ia-like antigens showed that liver and kidney contain amounts of Ia-like antigens comparable to spleen and lymphnode.[73,74] Ia-like antigens in detergent-solubilized cell membranes derived from kidney and liver are indistinguishable from Ia-like antigens derived from cultured B lymphoid cells as far as molecular structure and xeno- and alloantigenic properties are concerned.[75] Furthermore, testing of cell suspensions prepared by enzyme treatment of kidneys with a xenoantiserum in a rosette assay[76] and indirect immunofluorescence staining of frozen sections of kidney with a xenoantiserum[75] localized these antigens on endothelial cells of glomeruli. More recently, Scott et al.[77,78] tested sections of ethanol fixed paraffin-embedded tissue specimens from different parts of the human gastrointestinal tract with a rabbit antiserum to human Ia-like antigens in indirect immunofluorescence. These investigators found a specific patchy staining apically in the columnar cells of the normal small intestine decreasing in intensity from the top of the villi towards the crypts; on the other hand, Ia-like antigens could not be detected in

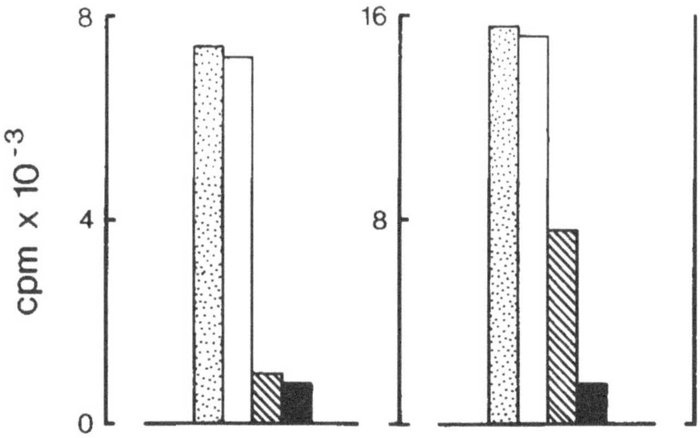

FIGURE 5. Effect of monoclonal antibodies to Ia-like antigens on the proliferative response of T lymphocytes to autologous and allogeneic monocyte-macrophages. Nylon wool purified T lymphocytes were incubated with irradiated autologous (left panel) and allogeneic (right panel) monocyte-macrophage in medium with fetal calf serum alone (▒) or in the presence of either anti HLA-A,B antigen monoclonal antibody Q1/28 (□) or anti Ia-like antigen monoclonal antibodies Q5/6 (▨) and Q5/13 (■).

colon and stomach epithelium. These results, in conjunction with the reports that in guinea pigs Ia-like antigens are expressed by epithelial cells lining the intestinal tract, the gall bladder, the respiratory tract, and the urinary tract, as well as by cells of parotid and of submandibular gland,[22,79–81] have raised the possibility that in man Ia-like antigens may have a wider tissue distribution than originally assumed. Therefore, we have tested a variety of surgically removed tissues with a monoclonal antibody to a framework determinant of Ia-like antigens in indirect immunofluorescence.[22,82] The results are summarized in Table 5 which also contains information about the tissue distribution of Ia-like antigens in monkeys, rabbits, and guinea pigs. Figures 6 and 7 show representative patterns of the staining of frozen sections of tissues with the monoclonal antibody Q5/13 to a common determinant of human Ia-like antigens.

Among tissues of ectodermal origin, Ia-like antigens are expressed by cells with the morphological and topographical features of astrocytes, but are not detectable on neurons.[82] Melanocytes, even those from pigmented areas of skin and those transformed into nevi cells, were not stained by the monoclonal antibody Q5/13 and by other monoclonal antibodies to distinct determinants of Ia-like antigens.[83] Mammary gland epithelial cells as well as endometrial cells were weakly stained with significant variations in the intensity of the staining among cells of a single gland and among samples removed from different subjects.[82] These differences may reflect the hormonal control of the expression of Ia-like antigens, since in guinea pigs and mice the phenotypic expression of these antigens on epithelial cells of mammary glands is increased by pregnancy and lactation and may be induced in virgin animals by lactotropic hormones.[84] Cells of the acinar portion of parotid glands were stained by the monoclonal antibody Q5/13.[22,82]

Among cells of the mesodermal-mesenchymal lineage, liver Kuppfer cells and skin Langerhans cells were strongly stained by the monoclonal antibody Q5/13.[22,82] The results with skin Langerhans and liver Kuppfer cells corroborrate previous findings obtained with conventional DR alloantisera and xenoantisera[19,85,86] and are in agreement

Table 5
DISTRIBUTION OF HUMAN, MONKEY, RABBIT, AND GUINEA PIG Ia-LIKE ANTIGENS IN NORMAL TISSUES OF NONLYMPHOID ORIGIN[a]

		Species			
Organ	Cell type	Man	Monkey	Rabbit	Guinea pig
Esophagus	Epithelium	−	−	−	−
Stomach	Epithelium	±	−	NT[b]	−
Small intestine	Epithelium	+	+	−	+
Colon	Epithelium	±	±	−	±
Sigma-rectum	Epithelium	−	−	−	−
Trachea	Epithelium	−	−	NT	−
Lung	Alveolar macrophages	+	+	+	−
Kidney	Glomerular capillaries	+	+	+	+
Kidney	Tubular capillaries	+	+	±	NT
Testis	Epithelium	−	−	−	−
Testis	Capillary endothelium	+	+	±	−
Uterus	Endometrium	+	−	−	−
Placenta	Endothelium	+	NT	NT	NT
Adrenal gland	Epithelium	−	−	−	−
Adrenal gland	Capillary endothelium	+	+	−	+
Liver	Kupffer	+	+	+	−
Mammary gland	Epithelium	±	−	−	−
Pancreas	Epithelium	−	−	−	−
Pancreas	Capillary endothelium	+	+	−	−
Parotid	Capillary endothelium	+	+	−	−
Parotid	Acinar cells	+	+	−	−
Skin	Langerhans	+	+	+	+
Thymus	Epithelium	+	+	+	±

[a]Determined by indirect immunofluorescence staining[22] of tissue cryostat sections with the monoclonal antibodies Q5/13 to human Ia-like antigens.
[b]Not tested.

with the ability of these cells to stimulate allogeneic lymphocytes[87,88] and to mediate antigen-specific T cell activation.[89,90] Cells of theca and of interstitial gland of ovary, as well as of endometrium and of epithelium of Fallopian tubes, express Ia-like antigens, but to a variable extent.[22,80] The expression of Ia-like antigens on endothelial cells varies from organ to organ and even within the same vessel: an interesting pattern is that detected in the renal vasculature.[91] In the renal cortex, the endothelial cells of arcuate and interlobular arteries are not stained by the monoclonal antibody Q5/13, while the endothelial cells of the intertubular capillary network show a strong reactivity with the monoclonal antibody Q5/13. In glomeruli, endothelial cells of arterioles are practically unreactive with the monoclonal antibody Q5/13 until they reach the point of entrance into glomeruli when a faint staining is detected; after arterioles branch out into glomerular tufts, endothelial cells are stained strongly. A similar bright staining was found in endothelial cells of glomerular capillaries and of efferent arteriole. In medulla, vasa recta and capillaries show a very strong fluorescent staining of endothelium. Our findings are in agreement with those obtained by Baldwin et al.[92] who also used a monoclonal antibody to a nonpolymorphic determinant of human Ia-like antigens, but differ from those of Hart et al.,[93] who, using similar reagents, detected Ia-like antigens on endothelial cells of large vessels. The reason for the conflicting results is not known; individual variations in the expression of Ia-like antigens cannot be ex-

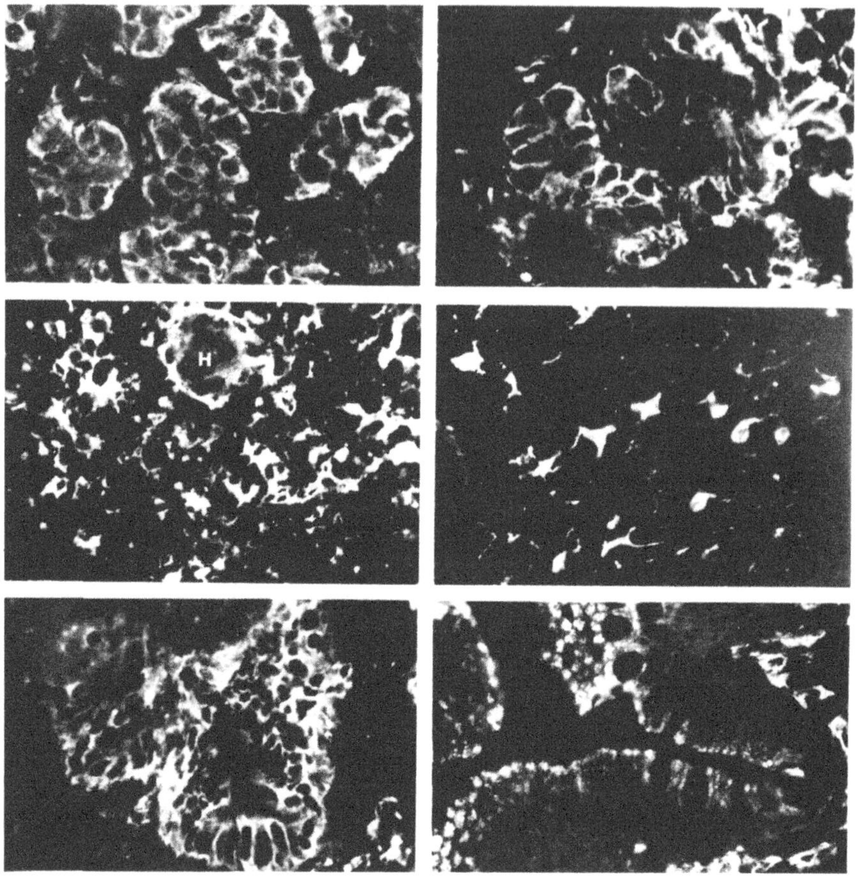

FIGURE 6. Expression of Ia-like antigens on normal human adult tissue. Four μ acetone-fixed cyrostat sections of surgically removed human tissues were reacted in indirect immunofluorescence with monoclonal antibody Q5/13 to human Ia-like antigens. Antral epithelial cells of stomach (upper left panel), acinar epithelial cells of parotid (upper right panel), dendritic cells of thymus medulla (middle left panel), Langerhans cells of epidermis (middle right panel), and cells of endometrium (lower left panel) show distinct linear fluorescence of the plasma membrane, while epithelial cells of jejunum (lower right panel) show a fine granular fluorescence of the apical portion (magnification × 650). H, indicate Hassal's body in some section planes.

cluded, since marked differences have been found in the content of histocompatibility antigens in a given tissue among different inbred rat strains.[94] The results of the studies with monoclonal antibodies altogether differ from those obtained with anti-DR alloantisera which usually stain weakly or not at all the substrates. It remains to be determined whether the conflicting results obtained with monoclonal antibodies and conventional alloantisera reflect differences in affinity of these two types of antibodies, differences in expression of common and allotypic determinants on Ia-like antigen bearing molecules synthesized by renal endothelial cells, or a cross-reaction of monoclonal antibodies with determinants of other antigens which are expressed in high amounts on renal endothelial cells. Expression of Ia-like antigens in the renal vasculature suggests that these antigens may be an important potential target for graft destruction by immunological mechanisms and explains why antibodies against Ia-like antigens may appear following kidney allografts mismatched for these antigens[95-97] and anti-donor B cell antibodies have been detected in eluates from rejected grafts.[98-100] Furthermore, one

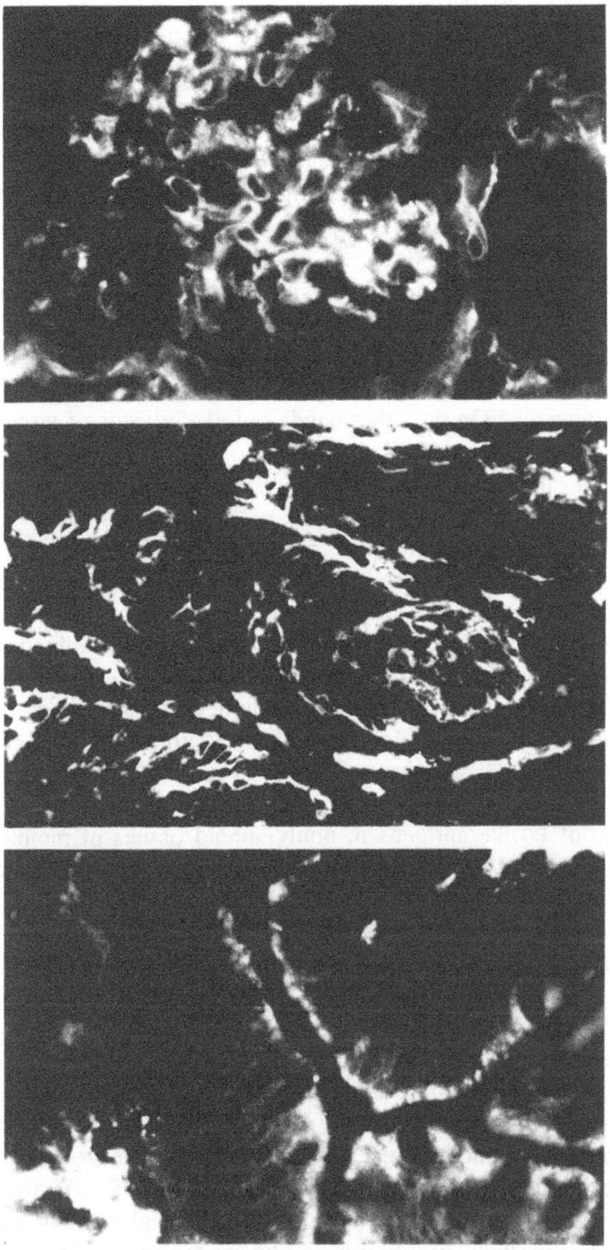

FIGURE 7. Reactivity of the monoclonal antibody Q5/13 to human Ia-like antigens with guinea pig and monkey adult tissues. Four μ acetone-fixed cryostat sections of tissues were reacted in indirect immunofluorescence with monoclonal antibody Q5/13. Glomerular endothelial cells of guinea pig kidney (upper panel; × 1000) and cells of monkey endometrium (middle panel; × 650) show distinct linear fluorescence of the plasma membrane, while epithelium cells of monkey jejunum (lower panel; × 650) show a fine speckled distribution in the apical portion.

has to reinterpret the specificity of antibodies detected in sera from patients with kidney allograft rejection and classified as directed to endothelial specific antigens since they do not react with HLA-A,B antigens as well as with heterophile antigens;[101,102] however, no data has been provided to exclude that these antibodies do not react with Ia-like antigens. It is also of interest to mention that not in all species Ia-like antigens are expressed on renal endothelial cells, as they have not been detected in rats[103] and pigs.[104] Therefore, conclusions drawn from studies in animal model systems on the role of antibodies to Ia-like antigens in kidney allograft rejection may not be valid for the human situation.

Ia-like antigens have been detected in endothelial cells of umbilical cord veins[105,106] and of capillaries in most organs, the most notable exceptions being brain cortex and placenta.[22,86] It is of interest that the endothelial cells of umbilical cord veins are able to stimulate allogeneic lymphocytes in mixed lymphocyte reactions[107] and to substitute for macrophages in the lymphoproliferative response of purified T cells to soluble protein antigens.[108] On the other hand, endothelial cells of large vessels do not appear to express Ia-like antigens.

Among tissues of entodermal derivation, Ia-like antigens are expressed on epithelial cells lining the gastrointestinal tract with the exception of esophagus, stomach, and rectum.[22,86] The level of Ia-like antigens, as judged by the intensity of immunofluorescence staining with polyclonal and monoclonal xenoantibodies, shows significant differences among epithelial cells located in different portions of the gastrointestinal tract, being highest in small intestine and in decreasing order in duodenum, jejunum, terminal ileum, and colon. In urinary bladder, only epithelial cells in deeper portions of crypts are weakly reactive with anti-Ia-like antigen xenoantibodies, while Ia-like antigens are not detectable in epithelial cells of kidney, ureter, and gall bladder. Finally, variable expression of Ia-like antigens was found in epithelial cells of bronchi and bronchial glands.

The distribution of Ia-like antigens in nonlymphoid tissues of monkeys and guinea pigs is superimposable to that in man (Table 5).[22] On the other hand, Ia-like antigens in rabbits have a more restricted tissue distribution than in other animal species, since these antigens are detected only in liver Kuppfer cells, thymic reticuloepithelial cells, alveolar macrophages, and endothelial cells of the capillary bed of colon, sigma, and rectum.[22]

In man, like in rabbits and guinea pigs, the full tissue distribution of Ia-like antigens is gradually attained during ontogeny: reticuloepithelial cells of thymus are the first to express detectable amounts of Ia-like antigens at the 20th week of gestation. The expression of these antigens is associated with the ability of these cells to stimulate allogeneic lymphocytes in mixed lymphocyte reactions. These reactions are specifically blocked by monoclonal antibodies to human Ia-like antigens (Figure 8). At a later time of intrauterine life, Ia-like antigens become detectable in renal glomerular and tubular capillaries. The complete tissue distribution of Ia-like antigens seen in adults is not reached until the 32nd week of gestation. Figure 9 shows representative patterns of the staining of frozen sections of tissues from 26-week old fetuses with the monoclonal antibody Q5/13 to human Ia-like antigens.

VIII. Ia-LIKE ANTIGENS ON TUMOR CELLS OF NONLYMPHOID ORIGIN

Malignant transformation of cells may be associated with changes in the expression of Ia-like antigens.[109–113] Epithelial cells of mammary glands which undergo malignant transformation may lose Ia-like antigens.[112,113] A variety of tumors of nonlympoid origin

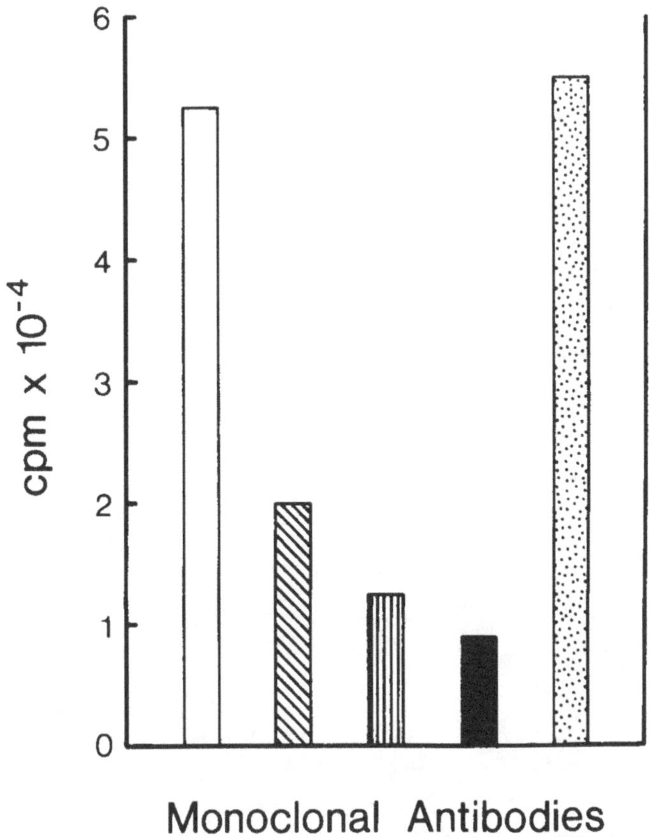

FIGURE 8. Effect of monoclonal antibodies to framework determinants of Ia-like antigens on the proliferative response of T lymphocytes to allogeneic fetal thymus cells. T lymphocytes purified by the nylon wool method from peripheral blood of the donor C.R. were incubated with irradiated cells obtained by mincing the thymus of a 26-week old fetus. The cultures were performed in medium with fetal calf serum (□) or in the presence of monoclonal antibody Q5/6 (▧), Q5/13 (▥), and CR-2 (■) to human Ia-like antigens and of monoclonal antibody Q1/28 (▦) to HLA-A,B antigens.

may acquire Ia-like antigens;[112,113] the phenomenon is especially frequent in melanoma and in other tumors of neurocrestal origin.[85,111,112] Ia-like antigens synthesized by melanoma cells have been the most extensively investigated for their immunological and structural properties: these antigens are immunologically functional since they can specifically bind antibodies in vitro,[109,110] can mediate complement and cell-dependent lysis of melanoma cells (Figure 10),[109,110,114] are immunogenic in xenogeneic combinations,[110] and can stimulate allogeneic lymphocytes in mixed lymphocyte reactions.[115,116] In a limited number of cases, the antigenic profile of Ia-like antigens bearing molecules on melanoma cells has been analyzed with conventional alloantisera which define the serological polymorphism of the DR system; when possible, the results have been compared with the DR phenotype of autologous B lymphocytes. Not more than two DR allospecificities have been detected on each melanoma cell line tested, and these specificities are genetically appropriate as they are in complete agreement with those found on autologous B lymphoid cells.[117–119]

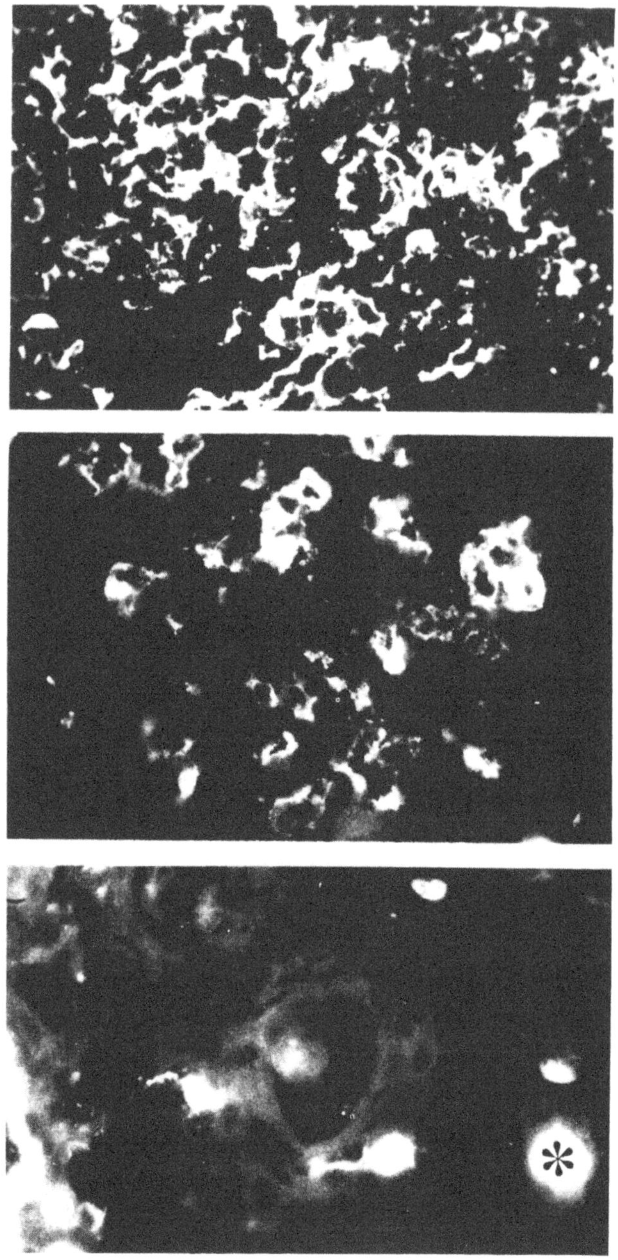

FIGURE 9. Expression of Ia-like antigens in fetal human cells of nonlymphoid origin. Four μ acetone-fixed cryostat sections of various tissues from a 26-week old human fetus were reacted in indirect immunofluorescence with monoclonal antibody Q5/13 to a framework determinant of human Ia-like antigens. Cortical dendritic cells of the thymus (upper panel; × 650), glomerular endothelial cells of the kidney (middle panel; × 1000), and dendritic cells of a hair follicle of the skin (lower panel; × 1000) show distinct linear fluorescence of the plasma membrane. ep. indicates epidermis; asterisk marks the cross-section of a hair.

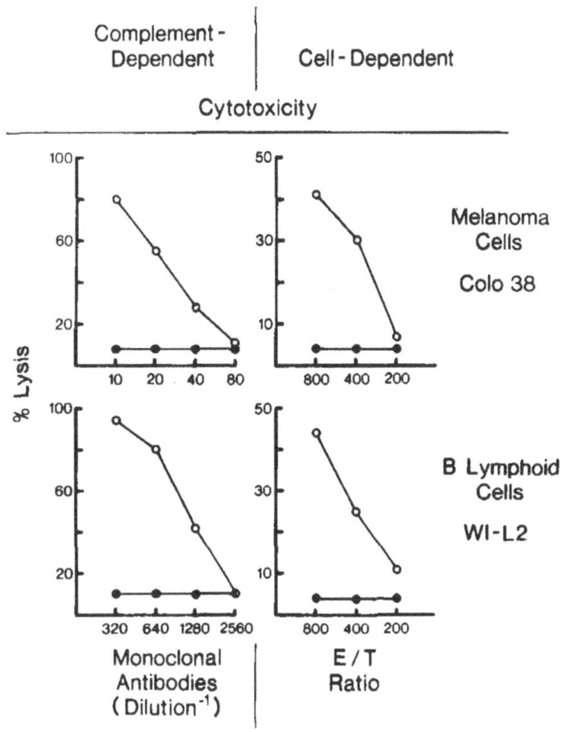

FIGURE 10. Lysis of melanoma cells Colo 38 and of B lymphoid cells WI-L2 by monoclonal antibody Q5/13 to human Ia-like antigens in conjunction with complement in the standard complement dependent cytotoxicity assay and with K cells in the antibody dependent cell mediated cytotoxicity assay.[53]

The appearance of Ia-like antigens on melanoma cells and other types of tumors is not restricted to cells in long-term culture and occurs also on surgically removed cells:[83] two lines of evidence argue against the possibility that Ia-like antigens on surgically removed cells are passively adsorbed from the milieu and not actively synthesized. First, freshly explanted melanoma cells continue to express Ia-like antigens when they are kept in short-term culture (Figure 11). Second, melanoma cells resynthesize Ia-like antigens when they are incubated at 37°C in tissue culture medium following removal of these antigens by digestion with pronase (Figure 11).

Cytofluorometric analysis of cultured melanoma cells stained with monoclonal antibodies to Ia-like antigens and fluorescein conjugated anti-mouse Ig antibodies has shown significant variations in the density of Ia-like antigens among various melanoma cell lines.[120] Intrinsic radiolabeling studies suggest that low expression of these antigens reflects reduced synthesis more than increased shedding of these antigens.[121] Melanoma cell lines as well as surgically removed melanoma lesions are quite heterogenous in the expression of Ia-like antigens. This heterogeneity does not appear to reflect cell cycle related changes in the expression of Ia-like antigens. Therefore, the relationship between cell cycle and expression of Ia-like antigens in melanoma is different from that in B lymphoid cells, since the expression of Ia-like antigens by cultured human B lymphoid cells WI-L2 is cell cycle dependent,[24] as discussed in another section of this chapter.

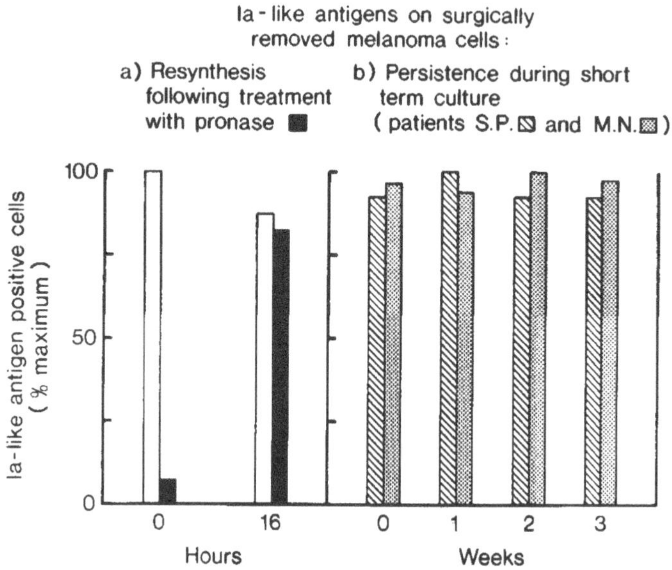

FIGURE 11. Expression of Ia-like antigens on surgically removed melanoma cells. Cell suspensions were prepared from surgically removed melanoma lesions. An aliquot of cells from patient P.I. was treated with pronase (30 min at 37°C). After removal of pronase, pronase-treated (■) and untreated (□) cells from patients P.I. were incubated at 37°C for 16 hr (left panel) and untreated cells obtained from patients S.P. (▧) and M.N. (▦) were incubated at 37°C up to 3 weeks (right panel) in medium containing 10% fetal calf serum. Expression of Ia-like antigens was determined in an indirect immunofluorescence assay.

Melanoma derived Ia-like antigens share several biological and structural properties with their B lymphoid cell counterparts: the level of Ia-like antigens on melanoma cells, like that on B lymphoid cells,[25,122] is reduced following incubation with inhibitors of protein synthesis,[113] but is not effected by incubation with interferon (Figure 12).[113,114] Ia-like antigens synthesized by melanoma cells, like B lymphoid cell derived antigens,[50,51] have a two-chain structure (Figure 4):[110,121] the apparent molecular weight of the two subunits shows small but reproducible differences among various melanoma cell lines but is in the range found for the two subunits of Ia-like antigens synthesized by B lymphoid cells. The differences in apparent molecular weight of the two subunits of Ia-like antigens are likely to reflect differences in the degree of glycosylation. Peptide mapping analysis and partial N-terminal sequencing of Ia-like antigens synthesized by the cultured melanoma cells SK-MEL-37 and by the autologous B lymphoblastoid cell line have not detected any significant difference in the primary structure of the polypeptide moiety of these antigens.[123] On the other hand, the degree of glycosylation of the two subunits of Ia-like antigens synthesized by melanoma cells may be different from that of the two subunits synthesized by B lymphoid cells. It is not known whether this different degree of glycosylation affects the immunological profile of Ia-like antigens. This possibility should be kept in mind since related studies have shown that the carbohydrate side chain of the heavy chain of HLA-A,B,C antigens may effect the expression of an antigenic determinant recognized by a monoclonal antibody.[124]

Ia-like antigens, like HLA-A,B, antigens,[125,126] are neither genetically linked nor spatially-structurally associated with serologically detectable tumor associated antigens identified in human melanoma cells with monoclonal antibodies and conventional

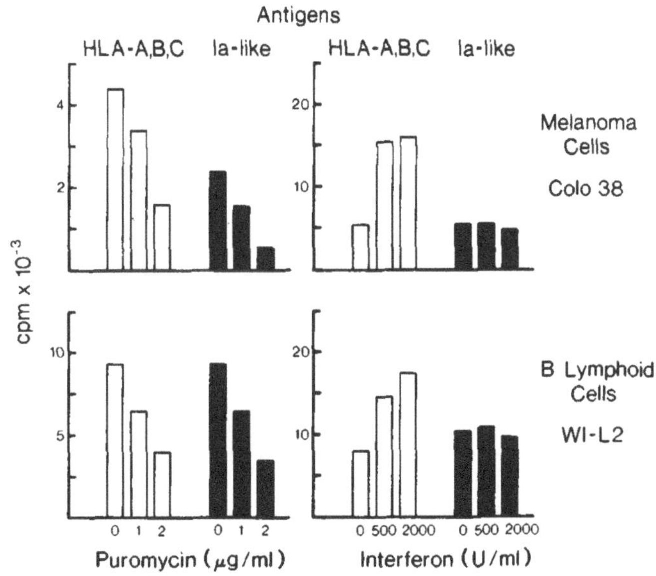

FIGURE 12. Effect of puromycin and interferon on the expression of Ia-like antigens on melanoma cells Colo38 and on B lymphoid cells WI-L2. Mid-log phase cells were incubated with puromycin and interferon at 37°C for 16 hr, washed 3 times and used as target cells in the ^{125}I-SpA radioimmunometric binding assay. Target cells (2.5×10^6) were incubated with 50 l of monoclonal antibodies W6/32 to a common determinant of HLA-A,B,C antigens or Q5/13 to a common determinant of Ia-like antigens and then reacted with ^{125}I-SpA (1×10^5 cpm).

xenoantisera.[127,128] The level of expression of Ia-like antigens on melanoma cells does not correlate with that of tumor associated antigens. Furthermore, analysis of the antigenic profile of melanoma cells moving through the growth cycle suggests that these two types of antigens are controlled by different regulatory mechanisms. However, Ia-like antigens may influence the immunogenicity of tumor associated antigens and the biology of tumor: melanoma associated antigens are less immunogenic in xenogenic combinations when expressed by melanoma cells which have lost Ia-like antigens following hybridization with rodent cells.[128] This finding parallels observations in guinea pigs that leukemia-specific antigens are immunogenic in syngeneic combinations only if expressed on cells which also carry Ia-like antigens.[129] Furthermore, in melanoma there is an inverse relationship between the percentage of Ia-like antigens bearing tumor cells and the degree of invasiveness of malignant lesions.[82] Finally, in breast carcinoma, Ia-like antigens are expressed by medullary carcinomas which have a favorable prognosis, but are not detectable in other types of breast carcinomas which have poor prognosis.[112,113]

IX. Ia-LIKE ANTIGENS IN BODY FLUIDS

Ia-like antigens, like HLA-A,B antigens,[130–132] are present in body fluids in an immunologically functional form, as they specifically bind antibodies in vitro and are immunogenic in allogeneic and xenogeneic combinations.[133–136] Thus antibodies to Ia-like antigens can be detected in sera from rabbits immunized with the lipoprotein fraction of human sera or with Ia-like antigens isolated from urine by binding to insolubilized anti-Ia-like antigen xenoantibodies. It is of interest that, in xenogeneic combi-

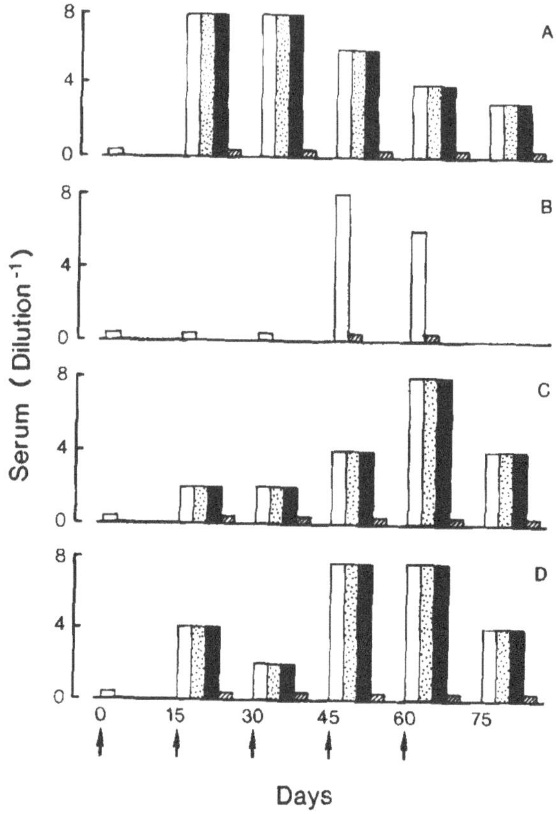

FIGURE 13. Analysis with the Fab_2 blocking assay of the specificity of cytotoxic antibodies developed by the uremic patients T.R. (panel A) and S.A. (panels B,C,D) immunized with plasma infusions from the donors B.R., R.A., F.A., B.U., and M.O. 1×10^5 cultured human B lymphoid cells WI-L2 (panel A), Daudi (panel B), Raji (panel C), and NC-37 (panel D) were incubated with 100 μl of Fab_2 fragments from normal rabbit serum (□), rabbit anti Ia-like antigen antiserum (⊡), rabbit anti human β_2-microglobulin (■), and a mixture (1 vol. and 1 vol.) of rabbit anti-Ia-like antigen antiserum and rabbit anti-human β_2-microglobulin (▨) for 60 min at room temperature and then used as targets in the cytotoxicity assay with serial bleedings from each patient. The arrows indicate immunizations.

nations, serum Ia-like antigens are less immunogenic than serum HLA-A,B,C antigens, while Ia-like antigens isolated from B lymphoid cells more immunogenic than HLA-A,B antigens isolated from cells.

Injection of cell free plasma may elicit cytotoxic antibodies to Ia-like antigens in patients without prior exposure to allogeneic histocompatibility antigens and boosts antibodies in patients immunized by blood transfusions. Representative results of studies performed in collaboration with Dr. U. Fagiolo are summarized in Figure 14 which shows the kinetic of formation of cytotoxic antibodies to Ia-like antigens in three patients immunized with plasma from selected donors.

The patients T.R. and S.A., who had no history of prior blood transfusions and pregnancies, developed cytotoxic antibodies to cultured B lymphoid cells following 2

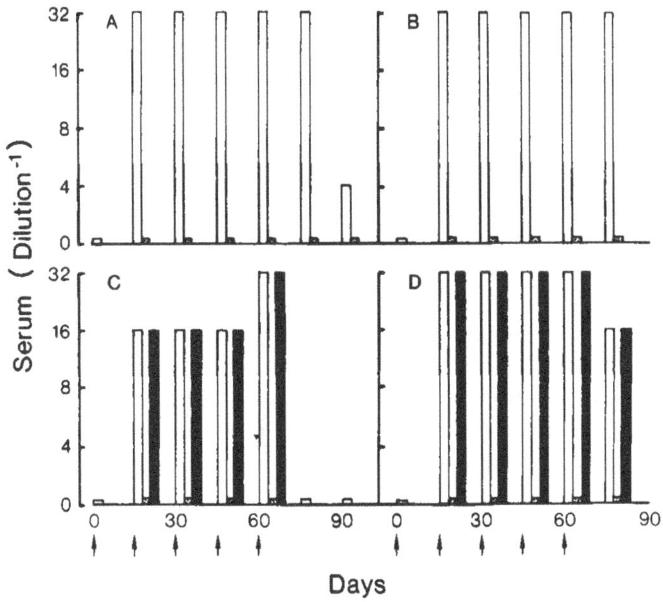

FIGURE 14. Analysis with the Fab_2 blocking assay of the specificity of cytotoxic antibodies developed by the uremic patient D.C. immunized with blood transfusions (A,C) and with plasma (B,D) from the donor T.O. 1×10^5 cultured human B lymphoid cells Daudi (A,B) and RPMI 8866 (C,D) were incubated with 100 μl of Fab_2 fragments from normal rabbit serum (□), rabbit anti-Ia-like antigen antiserum (⊡), and rabbit anti-human β_2-microglobulin (■) for 60 min at room temperature and then used as targets in the cytotoxicity assay with serial bleedings from the patient D.C. The arrows indicate immunizations.

injections of 50 ml of plasma from the donor B.R. The cytotoxicity was not blocked by coating the target cells with Fab_2 fragments from xenoantisera either to human β_2-microglobulin or to Ia-like antigens, but was inhibited by coating target cells with both types of Fab_2 fragment preparations (Figure 13). These results indicate that the antisera contain cytotoxic antibodies to Ia-like antigens and to β_2-μ associated structures, the latter being mostly HLA-A,B antigens. The patient D.C. had developed cytotoxic antibodies to Ia-like antigens following planned blood transfusions from the donor Z.W. When cytotoxic antibodies were not detectable, the patient was boosted with 50 ml of plasma from the same donor. Cytotoxic antibodies to cultured B lymphoid cells Daudi, RPMI 4098, RPMI 8866, and Raji reappeared in serum. Blocking experiments with Fab_2 fragments from an anti-human β_2-μ xenoantiserum, and from a xenoantiserum to human Ia-like antigens, showed that these antibodies are directed to Ia-like antigens (Figure 14).

The level of Ia-like antigens in body fluids is low and may increase in patients with chronic lymphocytic leukemia.[132] Because of the difficulties in isolating sufficient amounts of Ia-like antigens from body fluids, there has been limited characterization of their structure. According to our studies, Ia-like antigens in body fluids, like their cellular counterparts, are glycoproteins; on the other hand, Sandrin et al.[137] have identified a family of molecules with Ia-like antigenic activity which are glycolipids and have a molecular weight in the range of 2000 to 5000. The level of Ia-like antigen bearing glycolipids rises dramatically in normal subjects undergoing antigenic stimulation and is markedly reduced in patients with various types of malignant tumors.[138]

The mechanisms underlying these changes in the level of Ia-like antigens is not known. However, it has been suggested that measurement of these antigens may be useful for diagnosis and for following the clinical course of the disease in patients with malignancies.

X. CONCLUSION

The data we have reviewed indicate that the expression of Ia-like antigens is restricted to certain cell lineages and to particular differentiation stages within certain lineages. Furthermore, the appearance of Ia-like antigens can be induced by appropriate stimuli, some of which may be immunologic in nature, and may be associated with malignant transformation of cells. Ia-like antigens expressed by various types of cells and by malignant cells have a structure similar to that of Ia-like antigens synthesized by B lymphoid cells. The functional significance of the wide tissue distribution of Ia-like antigens is not fully understood. There is convincing evidence that expression of Ia-like antigens on certain cells of both lymphoid and nonlymphoid origin is associated with immune function: for instance, Ia-like antigens bearing lymphoid cells, macrophages, endothelial cells, and Langerhans cells have been shown to be potent stimulators in in vitro allogeneic mixed lymphocyte reactions and soluble-antigen presentation systems.[61,63,65,70–72,89,90] On the other hand, the role of Ia-like antigens on other cell types is still a matter of speculation; thus the expression of Ia-like antigens on hematopoietic stem cells and their disappearance with progression along the differentiation sequence has been suggested to reflect an involvement of these antigens in the control of hematopoietic differentiation and maturation.[17] It has been postulated that Ia-like antigens on reticuloepithelial cells of thymus may play a role in the immunological education of T cells,[138] and Ia-like antigens of epithelial cells of the gastrointestinal tract may be involved in the development of local immunity and in the homing of Ia-A-secreting plasma cells.[84]

Previous immunochemical studies had shown that the determinants defined by the monoclonal antibodies are expressed on subpopulations of human Ia-like antigens.[140,141] An interesting finding of the analysis of lymphoid cells with these monoclonal antibodies has been the heterogeneity among lymphoid cells in the expression of the determinants identified by the monoclonal antibodies. We do not know whether this heterogeneity is qualitative or quantitative, reflects a stage of maturation or differentiation, and has functional implications. These are areas of investigation we are pursuing using our library of monoclonal antibodies to human Ia-like antigens.

ACKNOWLEDGMENTS

This work was supported by the National Institutes of Health grants 19189, CA 32609, CA 32619, CA 32634, and CA 32635; by Progetto Finalizzato "Controllo della Crescita Neoplastica" CNR (Italy) No. 810132696, by a fellowship from A.I.R.C. (P.G.N.), and by a Research Career Development Award (M.A.P.). The authors wish to acknowledge the skilled secretarial assistance of Ms. Ellen Schmeding.

REFERENCES

1. **Moore, M. A. S., Broxmeyer, H. E., Sheridan, A. P. C., Meyers, P. A., Jacobsen, N., and Winchester, R. J.**, Continuous human bone marrow culture: Ia antigen characterization of probably pluripotential stem cells, *Blood*, 55, 682, 1980.
2. **Fitchen, J. H., Le Fevre, C., Ferrone, S., and Cline, M. J.**, Expression of Ia-like and HLA-A,B antigens on human multipotential hematopoietic progenitor cells, *Blood*, 59, 188, 1982.
3. **Cline, M. J. and Billing, R.**, Antigens expressed by human B lymphocytes and myeloid stem cells, *J. Exp. Med.*, 146, 1143, 1977.
4. **Janossy, G., Goldstone, A. H., Capellaro, D., Greaves, M. F., Kulenkampff, J., Pippard, M., and Welsh, K.**, Differentiation linked expression of p28,33 (Ia-like) structures on human leukaemic cells, *Br. J. Haematol.*, 37, 391, 1977.
5. **Winchester, R. J., Ross, G. D., Jarowski, C. I., Wang, C. Y., Halper, J., and Broxmeyer, H. E.**, Expression of Ia-like antigen molecules on human granulocytes during early phases of differentiation, *Proc. Natl. Acad. Sci. USA*, 74, 4012, 1977.
6. **Kaplan, J., Inoue, S., and Ottenbreit, J. J.**, Myeloid colony-forming cells express human B lymphocyte antigens, *Nature (London)*, 271, 458, 1978.
7. **Fitchen, J. H., Ferrone, S., Quaranta, V., Molinaro, G. A., and Cline, M. J.**, Moncolonal antibodies to HLA-A,B and Ia-like antigens inhibit colony formation by human myeloid progenitor cells, *J. Immunol.*, 125, 2004, 1980.
8. **Koeffler, H. P., Billing, R., Levine, A. M., and Golde, D. W.**, Ia antigen is a differentiation marker on human eosinophiles, *Blood*, 56, 11, 1980.
9. **Winchester, R. J., Meyers, P. A., Broxmeyer, H. E., Wang, C. Y., Moore, M. A. S., and Kunkel, H. G.**, Inhibition of human erythropoietic colony formation in culture by treatment with Ia antisera, *J. Exp. Med.*, 148, 613, 1978.
10. **Robinson, J., Sieff, C., Delia, D., Edwards, P. A. W., and Greaves, M.**, Expression of cell-surface HLA-DR, HLA-ABC and glycophorin during erythroid differentiation, *Nature (London)*, 289, 68, 1981.
11. **Belzer, M. B., Fitchen, J. H., Ferrone, S., Foon, K. A., Billing, R. J., and Golde, D. W.**, Expression of Ia-like antigens on human erythroid progenitor cells as determined by monoclonal antibodies and heteroantisera to Ia-like antigens, *Clin. Immunol. Immunopathol.*, 20, 111, 1981.
12. **Billing, R., Rafizadeh, B., Drew, I., Hartman, G., Gale, R., and Terasaki, P.**, Human B-lymphocyte antigens expressed by lymphocytic and myelocytic leukemia cells. I. Detection by rabbit antisera, *J. Exp. Med.*, 144, 167, 1976.
13. **Schlossman, S. F., Chess, L., Humphreys, R. E., and Strominger, J. L.**, Distribution of Ia-like molecules on the surface of normal and leukemic human cells, *Proc. Natl. Acad. Sci. USA*, 73, 1288, 1976.
14. **Winchester, R. J., Wang, C. Y., Halper, J., and Hoffman, T.**, Studies with B-cell allo- and hetero-antisera: parallel reactivity and special properties, *Scand. J. Immunol.*, 5, 745, 1976.
15. **Mohanakumar, T. and Raney, R. B., Jr.**, Leukaemia-associated antigens in man, *Clin. Haematol.*, 7, 363, 1978.
16. **Ross, G. D., Jarowski, C. I., Rabellino, E. M., and Winchester, R. J.**, The sequential appearance of Ia-like antigens and two different complement receptors during the maturation of human neutrophils, *J. Exp. Med.*, 147, 730, 1978.
17. **Winchester, R. J.**, Aspects of surface membrane differentiation during maturation of human hematopoietic and tumor cells: the Ia system, *Current Topics Dev. Biol.*, 14, 115, 1980.
18. **Rabellino, E. M., Nachman, R. L., Williams, N., Winchester, R. J., and Ross, G. D.**, Human megakaryocytes. I. Characterization of the membrane and cytoplasmic components of isolated marrow megakaryocytes, *J. Exp. Med.*, 149, 1273, 1979.
19. **Wernet, P.**, Human Ia-type alloantigens: methods of detection, aspects of chemistry and biology, markers for disease states, *Transplant. Rev.*, 30, 270, 1976.
20. **Quaranta, V., Indiveri, F., Glassy, M. C., Ng, A-K., Russo, C., Molinaro, G. A., Pellegrino, M. A., and Ferrone, S.**, Serological, functional and immunochemical characterization of a monoclonal antibody (MoAb Q2/70) to human Ia-like antigens, *Human Immunol.*, 1, 211, 1980.
21. **Russo, C., Indiveri, F., Quaranta, V., Molinaro, G. A., Pellegrino, M. A., and Ferrone, S.**, Use of monoclonal antibodies to investigate immunologic cross-reactivity of histocompatibility antigens from various animal species, *Transplant. Proc.*, 12, 376, 1980.
22. **Natali, P. G., Quaranta, V., Nicotra, M. R., Apollonj, C., Pellegrino, M. A., and Ferrone, S.**, Tissue distribution of Ia-like antigens in different species: Analysis with monoclonal antibodies, *Transplant. Proc.*, 13, 1026, 1981.

23. **Brickell, P. M., McConnell, I., Milstein, C., and Wright, B.**, A monoclonal antibody to the HLA-DR product recognizes a polymorphic Ia determinant in mice, *Immunology*, 43, 493, 1981.
24. **Sarkar, S., Glassy, M. C., Ferrone, S., and Jones, O. W.**, Cell cycle and the differential expression of HLA-A,B and HLA-DR antigens on human B lymphoid cells, *Proc. Natl. Acad. Sci. USA*, 77, 7297, 1980.
25. **Imai, K., Pellegrino, M. A., Ng, A-K., and Ferrone, S.**, Role of antigen density in immune lysis of interferon treated human lymphoid cells: analysis with monoclonal antibodies to the HLA-A,B antigenic molecular complex and to Ia-like antigens, *Scand. J. Immunol.*, in press.
26. **Greaves, M. F., Verbi, W., Festenstein, H., Papstteriadis, C., Jaraquemada, D., and Hayward, A.**, "Ia-like" antigens on human T cells, *Eur. J. Immunol.*, 9, 356, 1979.
27. **Yu, D. T. Y., Winchester, R. J., Fu, S. M., Gibofsky, A., Ko, H. S., and Kunkel, H. G.**, Peripheral blood Ia-positive T cells: increase in certain diseases and after immunization, *J. Exp. Med.*, 151, 91, 1980.
28. **Schurrman, R. K. B., Gelfand, K. B., Matheson, P., Zimmerman, B., and Posch, H. M.**, Identification of Ia on a subpopulation of human T lymphocytes that stimulate in a mixed lymphocyte reaction, *J. Immunol.*, 124, 1924, 1980.
29. **Evans, R. L., Faldetta, T. L., Humphreys, R. E., Pratt, D. M., Yunis, E. Y., and Schlossman, S. F.**, Peripheral human T cells sensitized in mixed leukocyte culture synthesize and express Ia-like antigens, *J. Exp. Med.*, 148, 1440, 1978.
30. **Reinherz, E. L., Kung, P. C., Pesaredo, J. R., Riz, J., Goldstein, G., and Schlossman, S. F.**, Ia determinants on human T cell subsets defined by monoclonal antibody. Activation stimuli required for expression, *J. Exp. Med.*, 150, 1472, 1979.
31. **Indiveri, F., Wilson, B. S., Russo, C., Quaranta, V., Pellegrino, M. A., and Ferrone, S.**, Ia-like antigens on human T lymphocytes: relationship to other surface markers, role in mixed lymphocyte reactions, and structural profile, *J. Immunol.*, 125, 2673, 1980.
32. **Fu. S. M., Chiorazzi, N., Wang, C. Y., Montazeri, G., Kunkel, H. G., Ko, H. S., and Gottlieb, A. B.**, Ia-bearing T lymphocytes in man. Their identification and role in the generation of allogeneic helper activity, *J. Exp. Med.*, 148, 1423, 1978.
33. **Johnsen, H. E., Madsen, M., and Kristensen, T.**, Lymphocyte subpopulations in man. Expression of HLA-DR determinants on human T cells in infectious mononucleosis, *Acta Pathol. Microbiol. Scand.*, 86, 307, 1978.
34. **Morito, T., Bankhurst, A. D., and Williams, R. C., Jr.**, Studies of T and B cell interactions in adult patients with combined immunodeficiency, *J. Clin. Invest.*, 65, 422, 1980.
35. **Tatsumi, E., Kimura, K., Takiuchi, Y., Fukuhara, S., Shirkawa, S., Uchino, H., Morikawa, S., and Minowada, J.**, T lymphocytes expressing human Ia-like antigens in infectious mononucleosis (IM), *Blood*, 50, 383, 1980.
36. **Williams, R. C., Jr., Webster, A. D. B., Morito, T., and Greaves, M. F.**, T lymphocyte subpopulations and Ia-positive T cells in patients with immunodeficiency, *Clin. Exp. Immunol.*, 42, 355, 1980.
37. **De Wolf, W. C., Schlossman, S. F., and Yunis, E. J.**, DRw antisera react with activated T cells, *J. Immunol.*, 122, 1780, 1979.
38. **Ko, H. S., Fu, S. M., Winchester, R. J., Yu, D. T. Y., and Kunkel, H. G.**, Ia determinants on stimulated human T lymphocytes. Occurrence on mitogen- and antigen-activated T cells, *J. Exp. Med.*, 150, 246, 1979.
39. **Metzgar, R. S., Bertoglio, J., Anderson, J. K., Bonnard, G. D., and Ruscetti, F. W.**, Detection of HLA-DRw (Ia-like) antigens on human T lymphocytes grown in tissue culture, *J. Immunol.*, 122, 949, 1979.
40. **Gillis, S., Baker, P. E., Ruscetti, F. W., and Smith, K. A.**, Long-term culture of human antigen-specific cytotoxic T-cell lines, *J. Exp. Med.*, 148, 1093, 1978.
41. **Suciu-Foca, N., Susinmo, E., Kiernam, P., Rohowsky, C., Weiner, J., and Rubinstein, P.**, DRw determinants on human T cells primed against allogeneic lymphocytes, *Transplant. Proc.*, 10, 845, 1978.
42. **Wilson, B. S., Indiveri, F., Quaranta, V., Pellegrino, M. A., and Ferrone, S.**, Level of Ia-like antigens on human B and T lymphocytes, *Transplant. Proc.*, 13, 1033, 1981.
43. **Dobloug, J. H., Forre, O., Lea, T., Solheim, B. G., and Natvig, J. B.**, Detection of Fc-receptor-bearing human lymphocytes. The majority of T cells carry HLA-DR antigens, *Clin. Immunol. Immunopath.*, 19, 224, 1981.
44. **Russo, C., Indiveri, F., Quaranta, V., Molinaro, G. A., Pellegrino, M. A., and Ferrone, S.**, Stimulation of human T lymphocytes by PHA-activated autologous T lymphocytes: analysis of the role of Ia-like antigens, *Immunogenetics*, 12, 267, 1981.
45. **Winchester, R. J. and Kunkel, H. G.**, The human Ia system, *Adv. Immunol.*, 28, 222, 1979.

46. **Samaraut, C., Gebuhrer, L., Brochier, J., Betuel, H., and Revillard, J. P.,** Presence of "Ia-like" antigens on human T lymphocytes bearing receptors for IgG, *Eur. J. Immunol.*, 7, 908, 1977.
47. **Palacios, R., Claesson, L., Moller, G., Peterson, P. A., and Moller, E.,** Role of HLA-DR antigens in the autologous mixed lymphocyte reaction. The invariant heavy chain but not the light polymorphic chain renders resting T cells sensitive to interleukin-2 and participates actively in the production of the T cell growth factor, *Eur. J. Immunol.*, in press.
48. **Indiveri, F., Scudeletti, M., Viglione, D., Guloni, V., Russo, C., Pellegrino, M. A., and Ferrone, S.,** Modulation of Ia-like antigens on PHA activated T lymphocytes, in *14th International Leucocyte Culture Conference, 1981,* Geelen, J., Ed., Elsevier/North-Holland Biomedical Press, Amsterdam, in press.
49. **Altovogt, P., Fohlman, J., Kurnick, J. T., Peterson, P., and Wigzell, H.,** Biochemical comparison of HLA-DR molecules derived from autologous human T and B lymphoblasts, *Eur. J. Immunol.*, 10, 908, 1980.
50. **Ferrone, S., Allison, J. P., and Pellegrino, M. A.,** Human DR (Ia-like) antigens: biological and molecular profile, *Contemp. Topics Molecul. Immunol.*, 7, 239, 1978.
51. **Allison, J. P., Walker, L. E., Russell, W. A., Pellegrino, M. A., Ferrone, S., Reisfeld, R. A., Frelinger, J. A., and Silver, J.,** Murine Ia and human DR antigens: homology of amino-terminal sequences, *Proc. Natl. Acad. Sci. USA*, 75, 3953, 1978.
52. **Niandet, P., Greaves, M., and Horwitz, D.,** Phenotypes of "null" lymphoid cells in human blood, *Scand. J. Immunol.*, 9, 387, 1979.
53. **Ng, A-K., Indiveri, F., Russo, C., Quaranta, V., Pellegrino, M. A., and Ferrone, S.,** Characterization of human null cells isolated from peripheral lymphocytes by a simultaneous double rosetting procedure, *Scand. J. Immunol.*, 14, 225, 1981.
54. **Hammerling, G. J., Deak, B. D., Mauve, G., Hammerling, U., and McDevitt, H. O.,** B lymphocyte alloantigens controlled by the I region of the major histocompatibility complex in mice, *Immunogenetics*, 68, 1974.
55. **Unanue, E. R., Dorf, M. E., David, C. S., and Benacerraf, B.,** The presence of I-region-associated antigens on B cells in molecules distinct from immunoglobulin and H-2K and H-2D, *Proc. Natl. Acad. Sci. USA*, 71, 5014, 1974.
56. **Delovitch, T. L. and McDevitt, H. O.,** Isolation and characterization of murine Ia antigens, *Immunogenetics*, 2, 39, 1975.
57. **Schwartz, R. H., Dickler, H. B., Sachs, D. H., and Schwartz, B. D.,** Studies of Ia antigens on murine peritoneal macrophages, *Scand. J. Immunol.*, 5, 731, 1976.
58. **Yamashita, U. and Shevach, E. M.,** The expression of Ia antigens on immunocompetent cells in the guinea pig. II. Ia antigens on macrophages, *J. Immunol.*, 119, 1584, 1977.
59. **Winchester, R. J., Fu, S. M., Wernet, P., Kunkel, H. G., Dupont, B., and Jersild, C.,** Recognition by pregnancy serums of non-HLA alloantigens selectively expressed on B lymphocytes, *J. Exp. Med.*, 141, 924, 1975.
60. **McKinney, T. K., Boto, W. O., Spiro, R. C., and Humphreys, R. E.,** Membrane expression and synthesis of p23,30 (HLA-DR antigen) by human peripheral blood monocytes, *Exp. Hematol.*, 8, 709, 1980.
61. **Hirschberg, H., Kaakinen, A., and Thorsby, E.,** Presence of HLA-D determinants on human macrophages, *Nature (London)*, 263, 1976.
62. **Drew, S. I., Nathanson, D. S., Zamfirescu, P., Carter, B. M., Terasaki, P. I., Naeim, F., and Billing, R.,** Serological characterization of human monocytes for HLA-B-lymphocyte, granulocyte and monocyte-associated antigens by cytotoxicity testing, *Tissue Antigens*, 11, 385, 1978.
63. **Raff, H. V., Picker, L. J., and Stobo, J. D.,** Macrophage heterogeneity in man. A subpopulation of HLA-DR-bearing macrophages required for antigen-induced T cell activation also contains stimulators for autologous-reactive T cells, *J. Exp. Med.*, 152, 581, 1980.
64. **Albrechtsen, D.,** HLA-D-associated "Ia-like" antigens on human macrophages, *Scand. J. Immunol.*, 6, 907, 1977.
65. **Ng, H-K., Zhang, Y., Russo, C., and Ferrone, S.,** Immunochemical and functional analysis of Ia-like antigens on human monocyte-macrophages with monoclonal antibodies, *Cell. Immunol.*, in press, 1982.
66. **Arenson, E. B., Jr., Epstein, M. B., and Seeger, R. C.,** Volumetric and functional heterogeneity of human monocytes, *J. Clin. Invest.*, 65, 613, 1980.
67. **Beller, D. I. and Unanue, E. R.,** Regulation of macraphage populations. II. Synthesis and expression of Ia antigens by peritoneal exudate macrophages is a transient event, *J. Immunol.*, 126, 263, 1981.
68. **Broxmeyer, H. E.,** Lactoferrin acts on Ia-like antigen-positive subpopulations of human monocytes to inhibit production of colony stimulatory activity in vitro, *J. Clin. Invest.*, 64, 1717, 1979.

69. **Schwartz, B. D., Paul, W. E., and Shevach, E. M.,** Guinea-pig Ia antigens: functional significance and chemical characterization. *Transplant. Rev.*, 30, 174, 1976.
70. **Thorsby, E., Bergholtz, B., and Nousiainen, H.,** Involvement of HLA cell-membrane molecules in T cell immune responses. Immunobiological and clinical significance, in *Current Trends in Histocompatibility*, Reisfeld, R. A. and Ferrone, S., Eds., Plenum Press, New York, 1981, 265.
71. **Gluckman, J. C. and Degoulet, P.,** Different stimulating capacity of monocytes and B lymphocytes in mixed leukocyte cultures: a dose response study, *Tissue Antigens*, 13, 278, 1979.
72. **Mann, D. L. and Abelson, L.,** Monocyte function in mixed lymphocyte reactions, *Cell. Immunol.*, 56, 357, 1980.
73. **Koyama, K., Nakamuro, K., Tanigaki, N., and Pressman, D.,** Alloantigens of human lymphoid cell lines; "human Ia-like antigens." Alloantigenic activity and cell line, organ and tissue distribution as determined by radioimmunoassay, *Immunology*, 33, 217, 1977.
74. **Williams, K. A., Hart, D. N. J., Fabre, J. W., and Morris, P. J.,** Distribution and quantitation of HLA-ABC and DR (Ia) antigens on human kidney and other tissues, *Transplantation*, 29, 274, 1980.
75. **Koyama, K., Fukunishi, T., Barcos, M., Tanigaki, N., and Pressman, D.,** Human Ia-like antigens in non-lymphoid organs, *Immunology*, 38, 333, 1979.
76. **Hayry, P., von Willebrand, E., and Andersson, L. C.,** Expression of HLA-ABC and -DR locus antigens on human kidney, endothelial, tubular and glomerular cells, *Scand. J. Immunol.*, 11, 303, 1980.
77. **Scott, H., Solheim, B. G., Brandtzaeg, P., and Thorsby, E.,** HLA-DR-like antigens in the epithelium of the human small intestine, *Scand. J. Immunol.*, 12, 77, 1980.
78. **Scott, H., Brandtzaeg, P., Solheim, B. G., and Thorsby, E.,** Relation between HLA-DR-like antigens and secretory component (SC) in jejunal epithelium of patients with coeliac disease or dermatitis herpetiformis, *Clin. Exp. Immunol.*, 44, 233, 1981.
79. **Wiman, K., Curman, B., Forsum, U., Klareskog, L., Malmnas-Tjernlund, U., Rask, L., Tragardh, L., and Peterson, P. A.,** Occurrence of Ia antigens on tissues of non-lymphoid origin, *Nature (London)*, 276, 711, 1978.
80. **Forsum, U., Klareskog, L., and Peterson, P. A.,** Distribution of Ia-antigen-like molecules of non-lymphoid tissues, *Scand. J. Immunol.*, 9, 343, 1979.
81. **Selby, W. S., Janossy, G., Goldstein, G., and Jewell, D. P.,** T lymphocyte subsets in human intestinal mucosa: the distribution and relationship to MHC-derived antigens, *Clin. Exp. Immunol.*, 44, 453, 1981.
82. **Natali, P. G., De Martino, C., Quaranta, V., Nicotra, M. R., Frezza, F., Pellegrino, M. A., and Ferrone, S.,** Expression of Ia-like antigens in normal human nonlymphoid tissues, *Transplantation*, 31, 75, 1981.
83. **Natali, P. G., Cordiali-Fei, P., Cavaliere, R., Di Filippo, F., Quaranta, V., Pellegrino, M. A., and Ferrone, S.,** Ia-like antigens on freshly explanted human melanoma, *Clin. Immunol. Immunopath.*, 19, 250, 1981.
84. **Klareskog, L., Forsum, U., and Peterson, P. A.,** Hormonal regulation of the expression of Ia antigens on mammary gland epithelium, *Eur. J. Immunol.*, 10, 958, 1980.
85. **Klareskog, L., Malmnas-Tjernlund, U., Forsum, U., and Peterson, P. A.,** Epidermal Langerhans cells express Ia antigens, *Nature (London)*, 268, 248, 1977.
86. **Rowden, G., Lewis, M. G., and Sullivan, A. K.,** Ia antigen expression on human epidermal Langerhans cells, *Nature (London)*, 268, 247, 1977.
87. **Levis, W. R. and Miller, A. E.,** Leukocyte/skin cultures as a measure of histocompatibility in man, *Lancet*, 2, 357, 1972.
88. **Hirschberg, H. and Thorsby, E.,** The presence of HLA-D determinants on human skin cells, *Transplantation*, 21, 343, 1976.
89. **Braathen, L. R. and Thorsby, E.,** Allo-activating and antigen-presenting capacity of human epidermal Langerhans cells, *Scand. J. Immunol.*, 11, 401, 1980.
90. **Braathen, L. R., Berle, E., Mobech-Hanssen, U., and Thorsby, E.,** Studies on human epidermal Langerhans cells. II. Activation of human T lymphocytes to Herpes Simplex virus, *Acta Derm. Venereol. (Stockholm)*, 60, 381, 1980.
91. **Natali, P. G., De Martino, C., Marcellini, M., Quaranta, V., and Ferrone, S.,** Expression of Ia-like antigens on the vasculature of human kidney, *Clin. Immunol. Immunopath.*, 20, 11, 1981.
92. **Baldwin, W. M., Claas, F. H. J., van Es. L. A., and van Rood, J. J.,** Distribution of endothelial-monocyte and HLA antigens on renal vascular endothelium, *Transplant. Proc.*, 13, 103, 1981.
93. **Hart, D. N. J., Fuggle, S. V., Williams, K. A., Fabre, J. W., Ting, A., and Morris, P. J.,** Localization of HLA-ABC and DR antigens in human kidney, *Transplantation*, 31, 428, 1981.
94. **Hart, D. N. J. and Fabre, J. W.,** Quantitative studies on the tissue distribution of Ia and SD antigens in the DA and Lewis rat stains, *Transplantation*, 27, 110, 1979.

95. **Ferrone, S., Halasz, N. A., Collins, G., and Pellegrino, M. A.,** Humoral immunity in kidney transplant recipients, *Transplantation,* 23, 113, 1977.
96. **Soulillou, J. P., Peyrat, M. A., and Guenal, J.,** Association between treatment-resistant kidney-allograft rejection and post-transplant appearance of antibodies to donor B-lymphocyte alloantigens, *Lancet,* 1, 354, 1978.
97. **Ting. A. and Morris, P. J.,** Development of donor-specific B lymphocyte antibodies after renal transplantation. No correlation with graft outcome, *Transplantation,* 28, 13, 1979.
98. **Mohanakumar, T., Giedlin, M., DuVall, C., Rhodes, C., Phibbs, M., Mendez, G., Kaplan, A. M., and Lee, H. M.,** B-lymphocyte-specific antibodies in human renal allografts, *Transplant. Proc.,* 11, 397, 1979.
99. **Lee, H. M., Waldrep, J. C., Mendez-Picon, G., and Mohanakumar, T.,** Antibodies eluted from rejected human renal allografts: specificity to B lymphocytes, monocytes, primary kidney cells, and endothelial cells, *Transplant. Proc.,* 13, 108, 1981.
100. **Nunez, G., McPhaul, J., and Stastny, P.,** Characterization of B-cell and monocyte antibodies in recipient sera and in eluates from rejected kidney allografts, *Transplant. Proc.,* 13, 111, 1981.
101. **Cerilli, J., Holliday, J. E., Fesperman, D. P., and Folger, M. R.,** Antivascular endothelial cell antibody—its role in transplantation, *Surgery,* 81, 132, 1977.
102. **Paul, L. C., van Es, E. A., van Rood, J. J., van Leeuwen, A., Brutel De La Riviere, G., and Graeff, J. De,** Antibodies directed against antigens on the endothelium of peritubular capillaries in patients with rejecting renal allografts, *Transplantation,* 27, 175, 1979.
103. **Hart, D. N. J. and Fabre, J. W.,** Major histocompatibility complex in rat kidney, ureter, and bladder. Localization with monoclonal antibodies and demonstration of Ia-positive dendritic cells, *Transplantation,* 31, 318, 1981.
104. **Pawlec, G., Davies, H. S., and Pearson, J. D.,** Stimulation of lymphocyte proliferation by non-lymphoid porcine tissue cells, *Tissue Antigens,* 14, 367, 1979.
105. **Moraes, J. R. and Stastny, P.,** A new antigen system expressed in human endothelial cells, *J. Clin. Invest.,* 60, 449, 1977.
106. **Hirschberg, H., Moen, T., and Thorsby, E.,** Complement and cell-mediated specific destruction of human endothelial cells treated with anti-DRw antisera, *Transplant. Proc.,* 11, 776, 1979.
107. **Hirschberg, H., Evensen, S. A., Henrikson, T., and Thorsby, E.,** The human mixed lymphocyte-endothelium culture interaction, *Transplantation,* 19, 495, 1975.
108. **Hirschberg, H., Bergh, O. J., and Thorsby, E.,** Antigen-presenting properties of human vascular endothelial cells, *J. Exp. Med.,* 152, 249s–255s, 1980.
109. **Winchester, R. J., Wang, C-Y., Gibofsky, A., Kunkel, H. G., Lloyd, K. O., and Old, L. J.,** Expression of Ia-like antigens on cultured human malignant melanoma cell lines, *Proc. Natl. Acad. Sci. USA,* 75, 6235, 1978.
110. **Wilson, B. S., Indiveri, F., Pellegrino, M. A., and Ferrone, S.,** DR (Ia-like) antigens on human melanoma cells: serological detection and immunochemical characterization, *J. Exp. Med.,* 149, 658, 1979.
111. **Howe, A. J., Seeger, R. C., Molinaro, G. A., and Ferrone, S.,** Analysis of human tumor cells for Ia-like antigens with monoclonal antibodies, *J. Natl. Cancer Inst.,* 66, 827, 1981.
112. **Natali, P. G., De Martino, C., Quaranta, V., Bigotti, A., Pellegrino, M. A., and Ferrone, S.,** Changes in Ia-like antigen expression on malignant human cells, *Immunogenetics,* 12, 409, 1981.
113. **Pellegrino, M. A., Natali, P. G., Ng, A. K., Imai, K., Russo, C., Bigotti, A., Wolf, P., and Ferrone, S.,** Unorthodox expression of Ia-like antigens on human tumor cells of non lymphoid origin: structural properties and biological significance, in *Recent Progress in Diagnostic Laboratory Immunology,* Nakamura, R. M., Ed., Masson Publishing, New York (in press), 1981.
114. **Imai, K., Ng, A. K., Glassy, M. C., and Ferrone, S.,** Differential effect of interferon on the expression of tumor associated antigens and histocompatibility antigens on human melanoma cells: Relationship to susceptibility to immune lysis mediated by monoclonal antibodies, *J. Immunol.,* in press.
115. **Nousiainen, H., Thorsby, E., and Ferrone, S.,** unpublished observations, 1981.
116. **Pollack, M. S.,** The detection of weak allogeneic stimulation by DR positive tumor cell lines, *Transplant. Proc.,* 13, 1947, 1981.
117. **Pollack, M. S., Heagney, S., and Fogh, J.,** HLA typing of cultured human tumor cell lines: the detection of genetically appropriate HLA-A,B,C and DR alloantigens, *Transplant. Proc.,* 12, 134, 1980.
118. **Pollack, M. S., Livingston, P. O., Fogh, J., Carey, T. E., Oettgen, H. F., and Dupont, B.,** Genetically appropriate expression of HLA and DR (IA) alloantigens on human lemanoma cell lines, *Tissue Antigens,* 15, 249, 1980.
119. **Pollack, M. S., Heagney, S. D., Livingston, P. O., and Fogh, J.,** HLA-A,B,C and DR alloantigen expression on forty-six cultured human tumor cell lines, *J. Natl. Cancer Inst.,* 66, 1003, 1981.

120. **Burchiel, S. W., Imai, K., Ferrone, S., and Warner, N. L.,** Heterogeneity of cell surface antigen expression of human melanoma cells: analysis with monoclonal antibodies and flow cytometry, in *Hybridoma in Cancer Diagnosis and Treatment*, Mitchell, M. and Oettgen, H., Eds., Raven Press, New York (in press), 1981.
121. **Wilson, B. S., Indiveri, F., Molinaro, G. A.,** Quaranta, V., and Ferrone, S., Characterization of DR antigens on cultured melanoma cells by using monoclonal antibodies, *Transplant. Proc.*, 12, 125, 1980.
122. **Ferrone, S., Pellegrino, M. A., Belvedere, M., Reisfeld, R. A., Curry, R., and Allison, J. P.,** Human B cell antigens: biological and immunogenic properties, in *Protides of Biological Fluids*, 25th Colloquium 1977, Peeters, H., Ed., Pergamon Press, Oxford, 1978, 645.
123. **Alexander, S., Lloyd, K. O., and Strominger, J. L.,** Characterization of the HLA-DR antigen of a continuous human melanoma cell line, submitted for publication.
124. **Wilson, B. S., Glassy, M. C., Quaranta, V., Ng, A. K., and Ferrone, S.,** Effect of tunicamycin on the assembly and antigenicity of HLA antigens: analysis with monoclonal antibodies, *Scand. J. Immunol.*, in press.
125. **Curry, R. A., Quaranta, V., Pellegrino, M. A., and Ferrone, S.,** Serologically detectable human melanoma associated antigens are not genetically linked to HLA-A and B antigens, *J. Immunol.*, 122, 2630, 1979.
126. **Curry, R., Quaranta, V., Wilson, B. S., McCabe, R. P., Natali, P. G., Pellegrino, M. A., and Ferrone, S.,** Expression of HLA antigens on cultured human melanoma cells; lack of association with melanoma associated antigens, in *Current Trends in Tumor Immunology*, Ferrone, S., Gorini, S., Herberman, R. B., and Reisfeld, R. A., Eds., Garland STPM Press, New York, 1979, 347.
127. **Imai, K. and Ferrone, S.,** Indirect rosette microassay to characterize human melanoma-associated antigens recognized by operationally specific xenoantisera, *Cancer Res.*, 40, 2252, 1980.
128. **Imai, K., Galloway, D. R., and Ferrone, S.,** Serological and immunochemical analysis of the specificity of the xenoantiserum 8986 elicited with hybrids between human melanoma cells and murine fibroblasts, *Cancer Res.*, 41, 1028, 1981.
129. **Forni, G., Shevach, D. H., and Green, I.,** Mutant lines of guinea pig L2C leukemia. I. Deletion of Ia alloantigens is associated with loss in immunogenicity of tumor-associated transplantation antigens, *J. Exp. Med.*, 143, 1067, 1975.
130. **Pellegrino, M. A., Ferrone, S., Pellegrino, M. A., Oh, S. K., and Reisfeld, R. A.,** Evaluation of two sources of soluble HLA antigens: platelets and serum, *Eur. J. Immunol.*, 4, 250, 1974.
131. **Ferrone, S., Mittal, K. K., Pellegrino, M. A., Allison, J. P., and Reisfeld, R. A.,** Immunogenicity of HLA antigens purified from serum, *Transplantation*, 23, 7, 1977.
132. **Indiveri, F., Fagiolo, U., Pellegrino, M. A., and Ferrone, S.,** Uremic patients immunized with planned blood transfusions: Detection of HLA antibodies with the cytotoxic test and an indirect rosette microassay, *Transplant. Proc.*, 11, 167, 1979.
133. **Wilson, B. S., Indiveri, F., Pellegrino, M. A., and Ferrone, S.,** Production and characterization of DR xenoantisera: use for detection of serum DR antigens, *J. Immunol.*, 122, 1967, 1979.
134. **Indiveri, F., Wilson, B. S., Natali, P. G., Pellegrino, M. A., and Ferrone, S.,** Immunogenicity of DR antigens isolated from body fluids, *Transplant. Proc.*, 12, 121, 1980.
135. **Pellegrino, M. A. and Ferrone, S.,** HLA and tumor-associated antigens in urine, in *Handbook of Cancer Immunology*, Waters, H., Ed., Garland STPM Press, New York, 1981, 105.
136. **Wilson, B. S., Ng, A-K., Quaranta, V., and Ferrone, S.,** HLA polyclonal and monoclonal xenoantibodies: production, characterization and application to the study of HLA antigens, in *Current Trends in Histocompatibility*, Reisfeld, R. A. and Ferrone, S., Eds., Plenum Press, New York, 1981, 307.
137. **Sandrin, M. S., Vaughan, H. A., McKenzie, I. F. C., Tait, B. D., and Parish, C. R.,** The human Ia system: definition and characterization by xenogeneic antisera, *Immunogenetics*, 8, 185, 1979.
138. Sandrin, M. S., Henning, M. M., Vaughan, H. A., McKenzie, I. F. C., and Parish, C. R., Serum Ia levels during tumor growth in mice and humans, *J. Natl. Cancer Inst.*, 66, 279, 1981.
139. **Zinkernagel, R. M., Callahan, G. N., Althage, A., Cooper, S., Klein, P. A., and Klein, J.,** the lymphoreticular system in triggering virus-plus-self-specific cytotoxic T-cells. Evidence for T-help, *J. Exp. Med.*, 147, 882, 1978.
140. **Quaranta, V., Pellegrino, M. A., and Ferrone, S.,** Serological and immunochemical characterization of the specificity of four monoclonal antibodies to distinct antigenic determinants expressed on subpopulations of human Ia-like antigens, *J. Immunol.*, 126, 548, 1981.
141. **Quaranta, V., Tanigaki, N., and Ferrone, S.,** Distribution of antigenic determinants recognized by three monoclonal antibodies (Q2/70, Q5/6 and Q5/13) on human Ia-like allo-antigens and on their subunits, *Immunogenetics*, 12, 175, 1981.
142. **Indiveri, F., Wilson, B. S., Pellegrino, M. A., and Ferrone, S.,** Detection of human histocompatibility antigens with an indirect rosette microassay, *J. Immunol. Methods*, 29, 101, 1979.

Chapter 6

HLA-D, HLA-DR, AND OTHER Ia-LIKE ANTIGENS IN MAN

Peter Stastny, Edward J. Ball, Gabriel Nunez, and Jun Okada

TABLE OF CONTENTS

I. INTRODUCTION

The region of the main histocompatibility complex (MHC) that codes for the human Ia-like antigens has been discovered relatively late and has presented a tremendous challenge. The Ia antigens have been difficult because of their limited expression in only certain kinds of cells, and their preponderant manifestation in functional and cellular assays, rather than being ubiquitous targets like the other common transplantation antigens.

The importance of the Ia-like antigens stems from their role in the genetic control of the immune response, their associations with a variety of diseases, and their probable role in organ transplantation. The study of the human Ia-like antigens has rapidly become one of the most active areas of contemporary biology.

However, the state of our understanding of the human Ia-like antigens can be appropriately described as primitive. Serologic methods allow the detection of one series of alloantigens, the HLA-DR antigens. In addition there are some broad "supertypic" antibodies and some antibodies that react with probable Ia-like antigens in monocytes, endothelial cells, and subsets of T lymphocytes. Everything else is based on cellular methods, especially primary and secondary mixed lymphocyte culture (MLC). The allospecificities defined by primary MLC with homozygous typing cells (HTC) receive the name of HLA-D antigens. The chromosomal region in which the corresponding genes are located is called the HLA-D region.

The purpose of the present review is to summarize what is presently known about the relationships of HLA-D and DR products, and other human Ia-like antigens, from available typing methods and genetic studies. Data from structural studies or functional investigations will be referred to only inasmuch such information may shed light on relationships, definition, or genetics of the antigens coded by the HLA-D region.

The cellular methods for typing for HLA-D were developed in 1973, and the efforts of the 1975 International Histocompatibility Workshop culminated with the recognition of the specificities Dw1 through Dw8. Although the specificities were upheld 2 years later with the discovery of the D-related (DR) antigens in close correspondence, many investigators have expressed the view that the mixed lymphocyte culture (MLC) determinants, detectable in primary MLC, appear to show greater complexity than can be adequately resolved with presently available methods. The analysis of secondary MLC using primed lymphocytes has been even more complicated, suggesting the existence of stimulating determinants coded by other HLA loci and possibly also by loci which are not associated with the HLA chromosomal region.

Since the HLA-DR specificities were defined, it was observed that they were closely associated with HLA-D. However, the association was never complete. This suggested to some investigators that D and DR were separate entities, probably coded by separate loci. Other investigators, however, were unwilling to make conclusions about the genetics of HLA-D and DR, in view of obvious difficulties in the interpretation of the data, and because little true genetic information was available.

Family studies showing recombination and indicating that the D and DR determinants segregate independently, would be needed to demonstrate that separate loci are involved. Such recombinations might be expected to be quite rare if the presumed loci are separated by a very short distance.

A further difficulty stems from lymphocyte stimulation due to other areas of the HLA chromosomal region. Such stimulation may give rise to "false negative" HLA-D assignments. Also, cells from some individuals may fail to respond in MLC, although they do not have the corresponding antigen, giving rise to "false positive" results. One must remember that the positive assignment is based on a negative response.

Moreover, the serologic data present complexities that need to be fully resolved. The specificity DRw6 is complicated and not fully understood. The supertypic specificities MT1, MT2, MT3, and MB1, MB2, MB3 can be interpreted in different ways and cause serologic confusion. It is not known whether they are located on the same, or on different molecules as the DR antigens. Additional complexities come from recently developed monoclonal antibodies against HLA-DR antigens. They suggest further heterogeneity, the possible existence of isotypes of DR antigens, and complications resulting from random association of polypeptide chains, giving rise to hybrid molecules.

II. TYPING FOR HLA-D BY MIXED LYMPHOCYTE CULTURE (MLC) WITH HOMOZYGOUS TYPING CELLS (HTC)

The fact that lymphocytes from HLA-identical siblings do not stimulate each other, and usually give results that are similar to the values obtained in autologous controls, indicate that most of the lymphocyte–stimulating activity, which is detectable in primary MLC, is coded by genes of the HLA chromosome region.[1] Exceptions to this rule were explainable as being the result of chromosomal recombination and led to the mapping of the main lymphocyte stimulating genes in a region centromeric to HLA-B.[2] The low MLR response, as an indication of identity at the main lymphocyte activating locus, was found, subsequently, to work also when the donor of the stimulating cells had inherited identical lymphocyte-activating determinants from both parents.[3] Such homozygotes were found not to stimulate any of the family members who had inherited the same chromosomes and also some unrelated individuals who apparently possessed similar lymphocyte–activating determinants.[4] These observations served as the basis for the use of homozygous typing cells (HTC) as reagents for typing for the lymphocyte activating determinants. Technology was developed for freezing the HTCs for repeated use in typing experiments locally and also for distribution to other investigators.

The first extensive international exchange of HTCs was carried out under the auspices of the Sixth International Histocompatibility Workshop, in 1975.[5] In this workshop, the reactions of typing cells against the panel were analyzed using two by two tables and clustering of correlation coefficients, much the same way as HLA serum reactions are usually analyzed. As a result of this analysis, eight determinants were identified and given numerical designations. The locus coding for these antigens was given the name HLA-D.

In the performance of the actual MLC typing experiments, results are usually expressed as relative responses[4] by the method of the 75th percentile.[6] It is customary, in the normalization of the data, to take into account both the ability of different responders to react and the variation in the ability of different stimulators to stimulate. The values obtained in this type of computation have sometimes been called "double normalized values."[7]

The basis of HTC typing is schematically illustrated in Figure 1. The mixed lymphocyte culture is made unidirectional by irradiation of the stimulating cells, which in the case of the example are homozygous Dw4. The first responder, which lacks this antigen, gives a relative response of around 100%. The other two responders shown, one heterozygous and the other homozygous Dw4, recognize the Dw4 determinant as very similar to self, and give low reactions with relative responses of about 20%.

These "typing reactions" are usually not entirely negative and their upper limits may vary somewhat in different experiments and with different HTCs. It is difficult, therefore, to establish definite cut-off levels and a variety of methods have been used to attempt to utilize objective criteria for deciding whether a given cell donor, producing

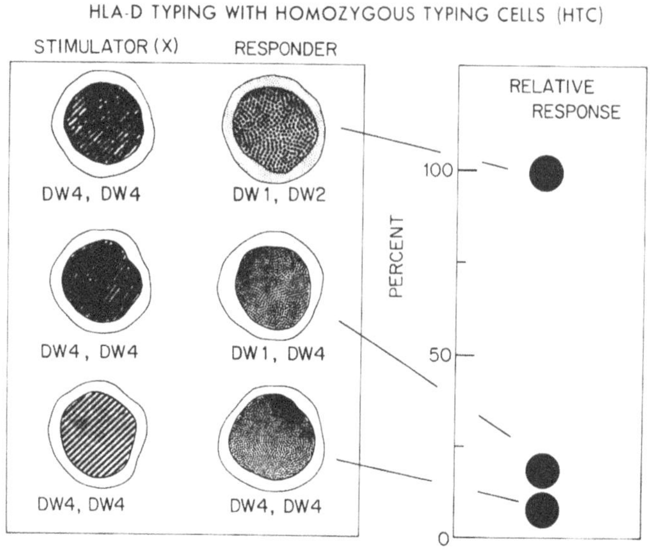

FIGURE 1. HLA-D typing with homozygous typing cells (HTC). Results of tritiated thymidine incorporation are expressed as relative responses (see text).

reactions which are not clearly negative or positive, should be considered to have the antigen in question. None of these methods appears to be entirely satisfactory and it seems best, in our experience, to consider some individuals to be untypable for technical reasons, rather than to make an entirely arbitrary assignment.

The cause for these intermediate reactions is not entirely clear. There are probably more than one. It is well known that X-irradiation, at the levels commonly used in these experiments, although adequate for suppressing DNA synthesis, may not stop other events associated with antigen recognition by the lymphocyte. Such cells have been found to be able to secrete lymphocyte activating factors, which increase the level of DNA synthesis in the responding cells. This phenomenon has been called "back stimulation"[8] and is observed when the responder is heterozygous and not when both cells are homozygous for the same determinant.

In addition, it is now generally recognized that the products of HLA-D are not the only lymphocyte stimulating determinants and therefore, it appears likely that intermediate reactions may result from the effect of stimulation by products of these other loci. For instance, it has been repeatedly observed that cells from sibs who have a recombination between HLA-A and B, and who are mismatched only in the HLA-A region, give rise to some lymphocyte stimulation to a level of approximately 25%.[9] It is not known whether the activation is due to HLA-A locus products themselves, or to products of an as yet not well-defined second mixed lymphocyte culture stimulating locus, situated in the proximity of HLA-A.

Recently, several groups of investigators have described HLA linked lymphocyte activating determinants which are detectable most clearly when the responding lymphocytes are first primed.[10–12] However, at least some of these lymphocyte–activating determinants, which are observed in the secondary reactions with primed cells, also have recently been observed to have an influence in primary mixed lymphocyte cultures.[13]

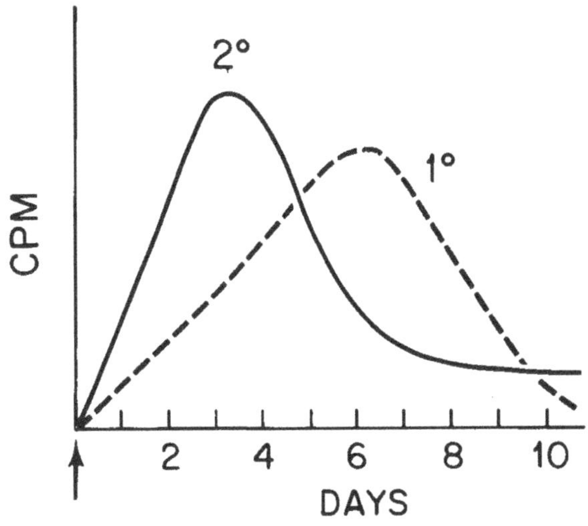

FIGURE 2. Primary and secondary responses in mixed lymphocyte cultures. The primary response usually peaks after 6 or 7 days of culture. The secondary peaks after 3 or 4 days.

III. TYPING FOR HLA-D BY SECONDARY MIXED LYMPHOCYTE CULTURE USING PRIMED LYMPHOCYTES

The secondary-type response to allogeneic antigens was studied by Andersen and Hayry[14] and subsequently by other investigators.[15] While the primary mixed lymphocyte culture proliferative response peaks at 6 or 7 days after initiation of culture, secondary cultures peak after only 3 or 4 days (Figure 2). This early response could be demonstrated by measuring the DNA synthesis in proliferating lymphocytes and also by cell mediated cytotoxicity against donor type target cells. The possibility of utilizing the secondary reaction as a method for typing for the allogeneic stimulating determinants was proposed by Sheehy and co-workers.[16] The method appeared attractive because it consisted of a positive assay to detect the presence of the antigens, and because the period of culture could be shortened to only 2 or 3 days. The idea was that primed lymphocytes would be frozen and utilized as reagents. Unfortunately, primed lymphocyte typing was found to be much more complicated than initially expected. At this writing, relatively few laboratories have succeeded to make this procedure a useful typing method.

There is little doubt that with certain selected primed cells it is possible to obtain reactions which coincide quite closely with the definition of the HLA-D determinants. More frequently, however, the secondary response appears to include reactions to HLA-DR[16] and to other antigens, revealing a complexity of responses which has been quite resistant to analysis.

One interpretation of the puzzling array of reactions obtained with primed lymphocytes is that the primary MLC is relatively simple because only the strongest antigens give rise to an important response. During priming, on the other hand, clones of lymphocytes capable of recognizing weaker activating determinants are expanded, and so, the primed lymphocyte method is capable of detecting a variety of weaker antigens. In a subsequent section we will discuss a series of antigens which are HLA linked, but distinct from HLA-D, and are detectable with primed lymphocytes.

IV. TYPING FOR THE HLA-D RELATED (DR) ANTIGENS

Many different investigators made preliminary observations which revealed the existence of a separate set of alloantibodies which reacted preferentially with B lymphocytes.[18–20] Such antibodies reacted with the Daudi cell line, known not to display any HLA-A, B, or C determinants.[18] They reacted with cells pretreated with turkey anti-human Beta 2 microglobulin,[21] and they were detectable in sera which had been subjected to exhaustive platelet absorption.[22] Preliminary work had suggested that such antibodies might have a relationship with HLA-D, because they were capable of inhibiting stimulation in mixed lymphocyte cultures.[23]

These efforts culminated in the Seventh International Histocompatibility Workshop[24] in 1977, with the definition of eight specificities which appear to be closely correlated with the first eight HLA-D specificities. The antigens were defined by classical serologic methods of analysis and their relationship with the HLA-D antigens was so striking that it was relatively easy to assign numbers to them that would correspond to those of the HLA-D alleles.

Standard methods for typing for these antigens were developed which consisted of isolating the B lymphocytes either physically, by rosetting of T cells, or by nylon wool filtration, or morphologically, by tagging them with an immunofluorescent marker as in the two-color immunofluorescence technique.[25] The typing reagents are alloantisera, which are usually platelet absorbed to remove antibodies against the HLA-A, B, C determinants. The method of detection most commonly used is a microcytotoxicity technique.

V. HLA-D/DR ANTIGENS IN DIFFERENT ETHNIC GROUPS

It is well known that HLA gene frequencies and relationships vary in different ethnic groups.[26] The major northern European haplotypes A1, B8, Dw3, DR3, and A3, B7, Dw2, DR2 decrease in frequency as one moves from north to south and they do not exist at all in Negro or in Mongoloid populations. It was not too surprising, therefore, to find that the very clear relationship between D and DR, in northern European populations, did not hold for some of the other ethnic groups.

Available information on this point comes mostly from studies performed in Japan,[27] and in indigenous populations of North[28–29] and South America.[30] In each of these studies, the serologically detectable HLA-DR antigens could be recognized with serologic patterns that were very similar to those obtained with cells of Caucasoid origin. The HLA-D determinants, as defined by using standard homozygous tying cells, were absent.[29]

However, new homozygous typing cells were found that appeared to identify determinants on Japanese cells, which appeared to take the place of the HLA-D antigens that were missing (Table 1). Thus the antigen DHO (now Dw12) associated with DR2 instead of Dw2; DYT took the place of Dw4; and the antigen DEn was found with DR blank in Japanese and appeared to be related to Dw6 in Caucasoids. Along similar lines, Layrisse and co-workers[30] have recently reported on HLA-D determinants which are characteristic of a population of Warao Indians in Venezuela.

The differential expression of HLA-D antigens is generally not complete. For example, Dw12 (DHO) is found also in Caucasoids, associated with DR2, like in the Japanese.[31] Thus DR2 associates with Dw2 in some Caucasoid haplotypes and with Dw12 in others.

Table 1
HLA-B, DR, D HAPLOTYPES IN JAPANESE AND CAUCASOID POPULATIONS

Japanese haplotypes	Caucasoid haplotypes
Bw52, DR2, Dw12[a]	B7, Dr2, Dw2
Bw54, DR4, DYT	Bw62, DR4, Dw4
Bw44, DR-, DEn	Bw38, DR6, Dw6.1

[a]Dw12 was previously called DHo.

Table 2
FREQUENCY OF HLA-D/DR CONCORDANT HAPLOTYPES

Ethnic group	Number tested	HLA-D/DR Concordant	
		Number	Percent
N.A. Caucasian	75	53	71
Am. Indian	30	14	47
Am. Black	17	4	24

VI. THE RELATIONSHIP BETWEEN HLA-D AND DR ANTIGENS

The close relationship between HLA-D and DR was an impressive finding when the DR antigens were first discovered. It led many investigators to postulate that the DR specificities were simply the serologic expression of the HLA-D antigens. However, it was a generalized experience, even in Caucasoid populations, that the correspondence was not perfect. There were always some cells that were positive for well-defined DR antigens, in which the corresponding HLA-D determinant was clearly absent.[28] In our own laboratory, these discordant D/DR typings were observed, both in the unrelated panel and in a number of carefully investigated families, typed with international workshop reagents. The frequency of HLA-D/DR concordant haplotypes in panel members of 3 ethnic groups, in our laboratory, is shown in Table 2.

To further investigate the significance of the apparent lack of agreement between HLA-DR and D typing in panel members, we investigated the effect of concordant and discordant typings on the results obtained in reciprocal MLC. The general conclusion of these investigations was that matching for HLA-DR was predictive for nonreactivity in MLC only if the HLA-D determinants were also matched.[32] An example of such an experiment is shown in Table 3. Individuals MS, BS, and BF had the phenotypes DR1, DR4. The first two were also Dw1 and Dw4, whereas in BF, Dw4 was missing. The results show that MS and BS were mutually nonstimulatory, but showed strong stimulation with BF. A summary of a number of such experiments is shown in Table 4. It shows that HLA-D matching had a profound effect on the reciprocal MLC, even when HLA-DR was always matched.

The serologic typings of these persons were carefully reviewed, looking for the possibility of subtle serologic differences. Even when more than 12 sera had been used to define the DR antigens, the HLA-D discrepant cells showed no differences in their

Table 3
MIXED LYMPHOCYTE CULTURE REACTIONS BETWEEN INDIVIDUALS HAVING HLA-DR1, DR4

		Stimulators		
Responders	HLA D,DR; D,DR	MS w1,1; w4,4	BS w1,1; w4,4	BF w1,1; -,4
MS	w1,1; w4,4	1	1	67
BS	w1,1; w4,4	5	3	72
BF	w1,1; --,4	63	60	1

Table 4
PREDICTIVE VALUE OF HLA-DR MATCHING FOR MLC NONREACTIVITY DEPENDS ON MATCHING FOR HLA-D

HLA-DR matched	HLA-D matched	Number	Result of MLC (DNV)
Yes	Yes	28	15.8 ± 6.7
Yes	No	33	72.3 ± 23.7[a]

[a]$p < 0.001$.

Table 5
INHERITANCE OF HLA ANTIGENS IN FAMILY 14

Family member	HLA genotypes	HLA haplotypes
Father	A2, B12, Cw5, DRw6, Dw6	A
	A11, B37, Cw6, DRw6, Dw6	B
Mother	Aw24,Bw44,Cw5,DR1, Dw1	C
	Aw24,B14, Cw2,DR1, D-	D
Sib 1		A,C
Sib 2		B,D
Sib 3		B,D
Sib 4		A,C

serologic reactions. These results only mean, of course, that with presently available reagents, no differences were found. Antibodies may be discovered in the future that may allow one to distinguish between these cells.

Similar observations have also been made in certain families. Of particular interest was family No. 14 (Table 5), in which both parents were homozygous for HLA-DR. The father was homozygous DRw6 and also homozygous Dw6. His cells did not stimulate any of the children and functioned as HTCs for Dw6, using unrelated responders. The mother is homozygous DR1. However, a typical Dw1 determinant is found in only one of her haplotypes. This haplotype was inherited by two of the children, whereas the other two children inherited a haplotype having DR1 and D blank. The cells from the mother stimulate all the children and they do not function as homozygous typing cells. Thus, this family illustrates the principle that homozygosity for HLA-DR does not predict homozygosity for HLA-D, and that the outcome of reciprocal stimulation depends on matching for HLA-D, even when HLA-DR is matched.

VII. CELL MEDIATED CYTOTOXICITY AGAINST HLA-D/DR ANTIGENS EXPRESSED IN MONOCYTES AND B LYMPHOCYTES

Prior to 1979, it was generally accepted that human cytotoxic lymphocytes did not recognize the HLA-D/DR determinants.[33-36] This was somewhat surprising, however, since in the mouse it is well established that cytotoxic cells against the I region determinants are easily detectable.[37] Feighery, in our laboratory,[38] found that by using stimulating cells matched to the responder for the HLA-A, B, C, antigens, and mismatched only for the HLA-D/DR region, and by using peripheral blood monocytes as target cells, cell mediated cytotoxicity against HLA-D was easily detectable. The determinants that serve as targets for this type of cytotoxicity, as demonstrated by cold cell inhibition experiments, were expressed in monocytes and B lymphocytes but not in T cells. Such cytotoxic cells recognized determinants closely related to the HLA-D/DR antigens in cells from unrelated donors and the cytotoxic reactions could be inhibited by xenoantibodies against human Ia antigens, alloantisera for the DR specificities, and more recently, with monoclonal antibody against monomorphic determinants of the HLA-DR molecules.[39-40] Ball and co-workers[41] recently observed, in a recombinant family, that the target determinants for cytotoxic lymphocytes of this type were encoded to the left of HLA-B.

It is possible also to generate cytotoxic lymphocytes against HLA-D/DR when the stimulating and responding cells are not completely matched for the HLA-A, B, C antigens, but then the target cells have to be carefully selected to avoid presence of HLA-A, B, C antigens against which the effectors might also be active. When this is done it can be shown that the magnitude of chromium release attributable to a single HLA-D/DR antigen is comparable to that obtained when the target is a single HLA-A or B determinant.[42]

When the effector cells are sensitized to stimulators carrying concordant HLA-D/DR antigens, it was observed that target cells lacking the HLA-D, but having the corresponding HLA-DR, were usually also killed and the amount of chromium released was similar.[38] An important question is whether such cytotoxic lymphocytes might also recognize the HLA-D antigen when the DR antigen is not present. Such reagents might then be used for typing for the HLA-D determinants by chromium release cytotoxicity. This would have many advantages over HTC typing. However, at this writing, it is not clear whether the HLA-D determinants can be separately recognized. It seems likely, however, that if properly stimulated, such clones of T cells can be developed. This would be an important new technical development, both for gaining a better understanding of the determinants that are coded by genes of the HLA-D region as well as offering the possibility of a practical typing procedure for application in clinical studies and for typing for organ transplantation.

VIII. ANTIGENS OTHER THAN HLA-D RECOGNIZED BY PRIMED LYMPHOCYTES

In a series of experiments designed to investigate the role of different HLA-D determinants in secondary allogeneic proliferation in man, Sasportes and co-workers[17] found that the HLA-DR determinants seemed to be more closely related to the stimulation of primed cells than HLA-D as defined with homozygous typing cells in primary cultures. Hartzman and co-workers[43] found that HLA-D and DR each were able to cause both primary and secondary stimulation. In their experiments, the responder and stimulator pairs were selected in such a way that only one of the two specificities was mismatched and therefore able to stimulate in the primary cultures.

More recently, several groups of investigators have provided evidence for the existence of still other HLA linked determinants capable of stimulating in primed lymphocyte tests. Mawas and co-workers[10] studied a family with a recombination between HLA-D/DR and GLO. Primed lymphocytes in this family gave indication of lymphocyte activating determinants separate from those of HLA-D/DR, located centromeric to the HLA-D/DR region. The antigens in this region to the left of D/DR, which they have called Beta, appear to produce weak primary and strong secondary MLR stimulation.

Termijtelen and co-workers[12] used priming between responding cells that were Dw3/DR3 positive and stimulating cells homozygous for Dw3/DR3. The lymphocytes primed in these experiments appeared to recognize a new determinant. It was different from the HLA-D/DR determinants, was associated with the A1, B8, Dw3, DR3 haplotypes and segregated with HLA in informative families. This antigen was called PL3A. Although it caused mainly stimulation in secondary cultures, it also appeared to play a role in the primary MLC.[13]

A somewhat similar approach was taken by Shaw and co-workers.[11] Extensive priming experiments were performed between subjects who were all phenotypically identical, having the antigens A1, A2, B7, B8, and most of them also matched for the antigens Dw2, Dw3, DR2, DR3. It consists of five specificities, each defined by at least two primed cells. The antigens were called "SB" for "secondary B cell antigens." These antigens appear to be independent and distinct from the D and DR antigens. They appear to be linked to HLA, in family studies, and in a recombinant were inherited with a portion of the chromosome centromeric to HLA-B. It is of interest that the SB determinants appear to be also targets for cell mediated cytotoxicity using lymphoblastoid B cells as target cells. However, no serum antibodies that correlated with these specificities were found.[44]

Thus the primed lymphocyte test seems suitable for the identification of a number of different lymphocyte activating determinants, some of which are distinct from HLA-D and DR. Recently, exchanges between Shaw, Termijtelen, and Mawas have been performed, giving preliminary evidence that PL3A, as well as the Dw3-associated PLT determinant found by Mawas, are similar to the SB1 antigen described by Shaw.

IX. OTHER Ia-LIKE ANTIGENS IN HUMAN B LYMPHOCYTES

We have seen that cellular methods have been used to identify some antigens other than the HLA-D/DR specificities which consitute B cell alloantigens that are presently not recognized using B cell typing sera. In addition, we have indications of the existence of serologically identifiable B cell alloantigens distinct from HLA-DR. For example, Mann and coworkers[45] performed a series of family studies with B cell alloantisera and obtained evidence of the existence of at least two loci for B cell antigens within the HLA region.

Serologic and immunochemical evidence for the existence of other distinct B cell alloantigens has been obtained by a number of different investigators. Tosi and coworkers[46] found that the antigen DC1 was a serologic specificity distinct from the DR antigens, which appeared to reside in separate molecules, as determined by immunoprecipitation studies.

The supertypic antigens MT1 (identical to DC1), MT2 and MT3 were extensively investigated during the Eighth International Histocompatibility Workshop.[47] Duquesnoy and co-workers[48] have evidence of yet another second supertypic system of antigens called MB antigens. The serologic analysis of these specificities poses difficulties because of the complexity of the sera that are used.

In addition to these experiments suggesting heterogeneity of B cell antigens on the basis of the epitopes used to characterize the allogeneic specificities, there appears to exist another kind of heterogeneity recognized by the use of monoclonal antibodies against framework determinants. Lampson and co-workers,[49] as well as Quaranta and co-workers,[50] reported that monoclonal antibodies, which apparently recognize backbone structures of the human DR antigens and react with B cells and monocytes from most individuals tested, appear to immunoprecipitate distinct molecules. Thus it appears that these monoclonal mouse anti-human DR antibodies are recognizing an isotype-like heterogeneity which is to a certain extent independent of the allospecificity.

Evidence for additional Ia-like molecules on human B cells is also obtained from the comparison of the reactivity of anti-p23/30 rabbit sera or chicken antibodies, and the monoclonal antibodies against the DR antigens. Particularly revealing in this regard have been studies performed by Nunez and co-workers,[51] using sera from kidney transplant recipients. From this work it appears that three kinds of B cell antibodies can be defined: (1) anti-DR antibodies, (2) antibodies against Ia-like antigens other than DR, and (3) antibodies against B cell antigens unrelated to the Ia-like antigens.

The last group of antibodies includes the cold reactive B cell lymphocytotoxins observed in the serum of many kidney transplant recipients and the B cell auto-antibodies which develop in patients with systemic lupus erythematosus. The reactions of these antibodies were not inhibited by pretreatment of the target cells with chicken anti-human Ia.

The anti-DR antibodies were defined by the fact that they were blocked when the B cells were pretreated by monoclonal antibody against monomorphic determinants of human DR antigens. Of particular interest was the second group of antibodies, which was defined by lack of blocking with the monoclonal anti-DR, but complete inhibition by pretreatment of the B cells with chicken anti-Ia. This group probably includes the broadly reactive antibodies MT1 and MT3, as well as other Ia-like specificities which have not yet been defined.

X. Ia-LIKE ANTIGENS OF HUMAN T CELLS AND T CELL SUBSETS

The majority of human T cells circulating in peripheral blood do not have detectable Ia-like antigens. However, a number of investigators have observed that when T cells are activated either under the influence of plant mitogens, such as PHA or concanavalin A, or by specific antigens, such as heterologous proteins or allogeneic antigens, the DR specificities become detectable on a substantial portion of the T lymphocytes undergoing blastic transformation.[52,53] Although binding of donor-type antigens to responding T cells is a possibility and has been observed in some experiments, it appears that the majority of the DR antigens expressed in such T cells are of the responder type and are synthesized by the cultured T cells. It has also been reported that in certain patients, who apparently are undergoing antigenic stimulation in vivo, T cells expressing the DR antigens can be found to be circulating in the blood.[54]

Recently Gazit and co-workers[55] have reported that human T cell blasts express some HLA-linked antigens which are distinct from the HLA-DR specificities.

It is well established from studies in mice that T cells express certain Ia specificities which are different from those expressed in B lymphocytes. These Ia antigens in the mouse have been found to be present in unstimulated T cells.

Van Leeuwen and co-workers[56] have recently reported on alloantisera that reacted

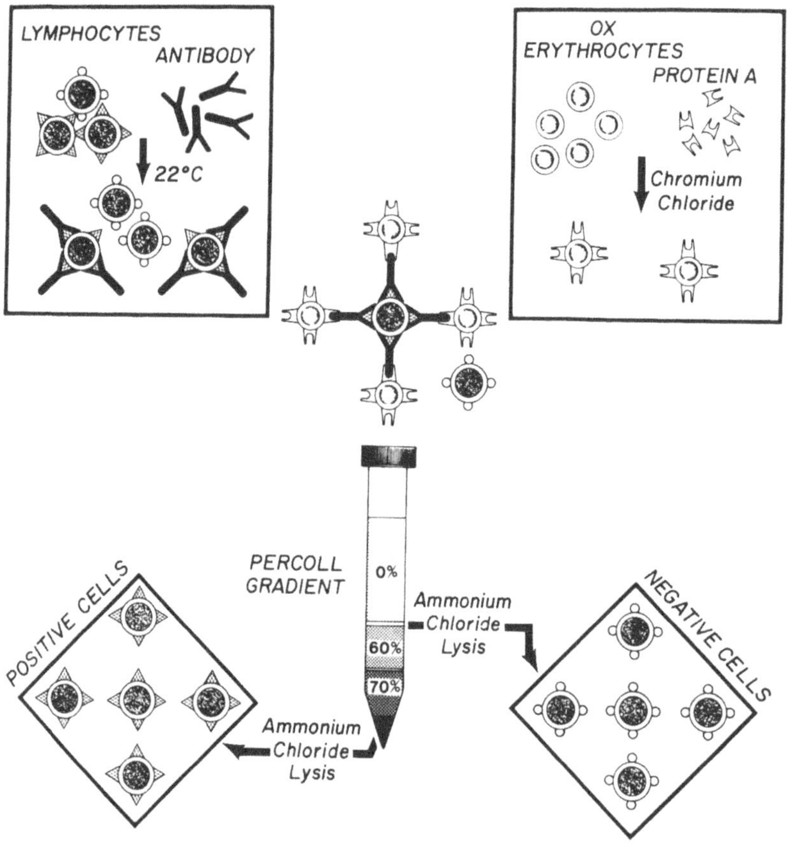

FIGURE 3. Method for the separation of T cell subsets using monoclonal antibodies and rosetting with staphylococcus protein A-coated ox erythrocytes.

with certain T cell subsets. Unstimulated T cells were fractionated on the basis of the presence of Fc receptors for IgG or IgM, and antibodies reacting with either the TG or the TM type of T lymphocytes were observed. The exact nature of these alloantibodies is not yet known. Many of them were not HLA linked.

In our laboratory, Okada has taken a somewhat different approach.[57] Normal resting T cells were separated into subsets by the use of monoclonal antibodies against the T cell phenotype markers OKT4 and OKT8. After combination of the antibody with the T cells carrying the marker, they were separated by rosetting with ox erythrocytes coated with staphylococcus protein A (Figure 3). Isolated T cell subsets were then utilized for screening alloantisera known to contain antibodies against B cells and monocytes, which had been platelet absorbed to remove HLA-A, B, C antibodies. In preliminary experiments, five sera reacting with the T8 subset and two reacting with the T4 subset were found. One serum, that gave strong reactions with the T8 subset of T lymphocytes, was of particular interest. This serum, when used to treat precursor lymphocytes with rabbit complement, markedly inhibited the development of supressor T cells, but had no substantial effect on the development of cytotoxic T cells. In preliminary experiments, the determinant expressed in the T8 subset, recognized by this pregnancy serum, appeared to be inherited in linkage with HLA. Preliminary results suggest that the determinants expressed in the T8 subset may be Ia-like and because of their presence in the precursors of supressor T cells but not of cytotoxic T cells, it is possible that the ZG antigens represent a human equivalent of the mouse I-J determinants.

XI. RECOMBINATION BETWEEN HLA-D AND HLA-DR ANTIGENS

Without inbred strains and without recombinants, progress in the study of the genetics of the HLA-D region has been slow compared to the work on the homologous region of the mouse. We have already presented some of the circumstantial evidence that suggests that HLA-D and HLA-DR can be considered distinct, but we have also emphasized the difficulties. We seem to be quite far from having a complete understanding of the molecular relationships and even further from figuring out the correspondence between genes and cell surface products. In view of the recent evidence that component polypeptide chains of the Ia molecules may be coded in separate genetic areas, and considering the recent evidence of recombinatorial interactions between alpha and beta chains, it is not easy to make inferences about genetic loci from the available data.

Several groups of investigators have published observations in families in which they believe a recombination between HLA-D and HLA-DR may have taken place.[58-60] Three of these will be briefly reviewed below.

Sachs and co-workers[59] have studied such a family quite extensively. The members were typed for HLA-A, B, C, D, DR, GLO, BF, C2, and C4 and several red cell markers. One member of the family, who should have been Dw7 homozygous, was found to behave at heterozygous. A possible explanation was that this individual inherited the Dw1 determinant present in the other paternal haplotype through a chromosomal recombination. However, the cells did not behave like the other Dw1 positive cells in the family. Several explanations were offered by the authors to explain the anomalous findings. Their favorite hypothesis is that Dw1 is complex, perhaps composed of two lymphocyte activiting determinants, and that only one of them was inherited by the anomolous family member. Other alternative explanations discussed are the possibility of a mutation in the HLA-D region and the possibility of development of hybrid antigens, products of gene complementation.

Another puzzling family was extensively investigated by Reinsmoen and co-workers.[58] The family included two siblings, who were HLA-A, B, C, and DR identical, but strongly stimulatory in mixed lymphocyte cultures. The hypothesis favored by these authors is that there are two separate, but closely linked loci, HLA-D and HLA-DR, both of which stimulate strongly in primary MLC. Furthermore, the HLA-D region appears to confer the specificity of the early reactions observed in secondary MLC. Other possibilities considered were a spontaneous mutation, detectable by MLC testing but not by serology, and the possibility of stimulation by determinants not coded by genes of the HLA region.

Suciu-Foca and co-workers[60] have studied a family in which one of the children inherited from the father a haplotype with DR7 and Dw3. In reciprocal primary mixed lymphocyte cultures, this child was MLC nonstimulatory with a DR3, Dw3, DR5, Dw5 sister, and was said to be MLC-different from the DR7, Dw7, DR5, Dw5 brother. In this family also, the secondary MLCs appeared to be activated by the DR antigens. Thus an interesting discrepancy resulted, since the secondary lymphocyte test did not show the anomolous reactions observed in primary MLC. The authors believe that HLA-D and DR antigens are coded by closely linked but distinct loci, that HLA-D antigens activate the recognition response in MLC, and DR antigens stimulate the response of primed T cells.

As already pointed out, these interesting observations must be interpreted with caution. The D region may be more complex than revealed by the discrimination of currently available testing techniques. Study of such families, with unusual typing results is undoubtedly useful and important. But we must await further developments before we can make difinitive interpretations.

XII. DISCUSSION

Although striking progress has been made in just a few years, we are still far from having a complete understanding of the HLA-D region antigens. The parallel development of cellular and serologic typing methods for the products of these genes has given rise to peculiar problems not encountered in the study of the other HLA antigens which are commonly defined by serologic methods only.

The idea that Class I HLA products can be characterized only by serologic techniques and the Class II products only by cellular methods was initially attractive, but turned out to be incorrect, and has really not been a useful framework of reference for thinking about these antigens. It is clear that the Class I and Class II type histocompatibility antigens are structurally and anatomically distinct. There are also functional differences. But they are not as crisp and definite as was initially envisioned.

The fact that human cytotoxic lymphocytes can recognize both Class I and Class II specificities is now well established. However, there is recent evidence that the phenotype of the effector cells may be different.[42] Not only are the HLA-DR antigens recognized by cytotoxic lymphocytes, but they also appear to be important in stimulation of the proliferation in secondary mixed lymphocyte cultures.[17]

It is amazing, in retrospect, how long it took to resolve the separation between the HLA-A and B loci, and how difficult it was to clarify the separate nature of the mixed lymphocyte culture stimulating determinants and the recognition of HLA-D. One should therefore not be discouraged if progress in understanding the HLA-D/DR antigens seems slow.

The data from the early mixed lymphocyte culture work has clearly shaped our view of the D region, but one must be cautiously skeptical about the picture derived from mixed lymphocyte cultures. Even if one makes a generous allowance for technical difficulties, there is a lack of correspondence between the results of primary and secondary cultures and it takes some work to fit the results obtained by HTC typing to a simple genetic hypothesis. On the other hand, it is interesting to note that the results obtained in Rhesus monkeys and chimpanzees are very similar to those obtained in man.[61]

Comparison of the HLA-D region with the homologous region of the mouse has been interesting and stimulating, but must be qualified by the fact that, in spite of the many similarities, there are also some major differences in the structure of the H-2 region of the mouse and the human HLA chromosomal complex. There are recent data indicating that, structurally, the molecules immunoprecipitated by antibodies against the DR antigens and those produced in mouse strains by immunization across an I-E difference are strikingly similar in their amino acid sequence. However, since it has recently been shown that the beta chain of these molecules, which is the polypeptide chain that appears to carry the allogeneic determinants, is coded in the mouse I-A region, it is possible that the narrow allogeneic specificities of the HLA-DR region actually correspond to mouse I-A.

The division of the mouse I region into several subregions that are associated with different immunoregulatory functions, code for MLC stimulation, and cause graft rejection guides some of the newer efforts in defining HLA-D specificities in man.

It appears now that available methods have yielded about as much information as is possible and that further progress will depend on the utilization of new approaches.

Among the new techniques that promise to further our understanding of the products and genetics of the HLA-D region, one should mention the investigation of Ia-like antigens expressed only in certain subsets of cells, the analysis of D region products with a multitude of monoclonal antibodies, and the development of T cell clones against HLA-D region products.

Examples of the first approach are the development of reagents which recognize determinants which are uniquely expressed in endothelial cells or monocytes or subsets of T lymphocytes. Some of the endothelial-monocyte (EM) antigens are HLA-linked and may well be Ia-like.[62] The recently discovered ZG antigens, which are expressed in the precursors of supressor T cells, and also in peripheral blood monocytes, could well represent the human equivalent of the mouse I-J antigens.[57]

The use of monoclonal antibodies will undoubtedly revolutionize the definition of the D region products. They have already begun to reveal a complexity that was previously unsuspected.[49,50] Further analysis of the D region products using such antibodies promises to be very informative. They should clarify the interactions between different alpha and beta chains, the possible existence of isotypes, and the different relationships between the allogeneic determinants and the sites recognized by the monoclonal antibodies. If the existence of isotypes is confirmed, it will be of interest to determine their anatomic distribution and expression in different types of cells and the possibility that different isotypes may have different functions.

Both proliferating and cytotoxic T cells can now be raised against HLA-D region products. There is preliminary evidence at least, that both types of T cells can respond to HLA-D and to HLA-DR antigens. Many laboratories are presently devoting efforts to the development of techniques for isolating and expanding clones of T cells to perform an analysis of the HLA-D region antigens.

Cytotoxic assays appear to offer some advantages over proliferative assays. They can be performed in a matter of hours instead of days, and they seem to have better reproducibility.

Immunization in vitro is a viable alternative to in vivo immunization and appears to work in many instances. If the responder and the stimulator are properly selected, usually matched for a number of known determinants and mismatched for the antigens of interest, and if the target cells used express the antigen to be determined, then this approach can be successful. It should be possible to begin to develop batteries of cytotoxic T cells recognizing many different HLA-D specificities. If this works, it is not difficult to imagine that assays based on cytotoxicity could replace the tests based on lymphocyte proliferation. Cellular methods are obviously here to stay for some time.

Moreover, it is possible that the development of human hybridomas, secreting specific antibodies, will follow closely, and that eventually it will be possible, also using in vitro planned immunization, to develop monoclonal antibodies against determinants recognized now only by T cells.

VIII. CONCLUSION

The human Ia-like antigens determine the collaboration between T cells and accesory cells that present antigen, they appear to play a role in the genetic control of immune responses, similar to that of the homologous systems in other species; they have been found to be markers of genetic predisposition in a variety of human diseases, and they appear to play a role in organ transplantation.

Understanding of the HLA-D region and its products has been colored by the peculiar combination of cellular and serologic methods that has been used in their study. Efforts to correlate these various methodologies have sometimes yielded puzzling results.

At the present time we recognize strong lymphocyte activating products detectable in primary MLC and responsible for the HTC typing reactions (D antigens), weaker lymphocyte activating products detectable mainly in secondary MLC, narrow serologic allospecifities (the DR antigens), broad alloantigens (the MT and MB series), and the new specificities recognized by monoclonal antibodies.

There is intense activity in many laboratories for development of new methodologies for HLA-D region studies. A number of new techniques are already on the horizon. They include the investigation of determinants expressed uniquely in certain types of cells such as endothelial cells, monocytes, or subsets of T lymphocytes; the development of monoclonal antibodies, the use of cytotoxic T cells, and the development of T cell clones against the HLA-D region products. As these methods are perfected, a large amount of new information is to be expected and many of the present difficulties may be resolved.

ACKNOWLEDGMENTS

Supported in part by NIH grants R01 AI12563 and R01 Am22185.

REFERENCES

1. **Dupont, B. and Hansen, J. A.,** Human mixed-lymphocyte culture reaction: genetics, specificity, and biological implications. *Adv. Immunol.*, 23, 107, 1976.
2. **Yunis, E. J. and Amos, D. B.,** Three closely linked genetic systems relevant to transplantation, *Proc. Nat. Acad. Sci. USA*, 68, 3031, 1971.
3. **Van den Tweel, J. G., Blusse van Oud Alblas, A., Keuning, J. J., Goulmy, E., Termijtelen A., Bach M. L., and van Rood, J. J.,** Typing for MLC (LD). I. Lymphocytes from cousin-marriage offspring as typing cells, *Transplant Proc.*, 5, 1535, 1973.
4. **Jorgensen, F., Lamm, L. V., and Kissmeyer-Nielsen, F.,** Mixed lymphocyte cultures with inbred individuals: an approach to MLC typing, *Tissue Antigens*, 3, 323, 1973.
5. **Thorsby, E. and Piazza, A.,** Joint report from the Sixth International Histocompatibility Workshop Conference. II. Typing for HLA-D (LD-1 or MLC) determinants, in: *Histocompatibility Testing 1975*, F. Kissmeyer-Nielsen, Ed., Munksgaard, Copenhagen, 1975, 414.
6. **Ryder, L. P., Thomsen, M., Platz, P., and Svejgaard, A.,** Data reduction in LD-typing, in *Histocompatibility Testing 1975*, F. Kissmeyer-Nielsen, Ed., Munksgaard, Copenhagen, 1975, 557.
7. **Festenstein, H. and Oliver, R. T. D.,** Cellular typing, in *Histocompatibility Testing 1977*, Bodmer, W. F., Batchelor, J. R., Bodmer, J. G., Festenstein, H., and Morris, P. J., Eds., Munksgaard, Copenhagen, 1978, 85.
8. **Sasazuki, T., McMichael, A., Radvany, R., Payne, R., and McDevitt, H.,** Use of high dose X-irradiation to block back stimulation in the MLC reaction, *Tissue Antigens*, 7, 91, 1976.
9. **Bijnen, A. B., Schreuder, I., Volkers, W. S., Parlevliet, J., and van Rood, J. J.,** The lymphocyte activating influence of the HLA-A region, *J. Immunogenetics*, 4, 1, 1977.
10. **Mawas, C., Charmot, D., and Mercier, P.,** Split of HLA-D into two regions and by a recombination between HLA-D and GLO. I. Study in a family and primed lymphocyte typing for determinants coded by the region, *Tissue Antigens*, 15, 458, 1980.
11. **Shaw, S., Johnson, A. H., and Shearer, G. M.,** Evidence for a new segregant series of B cell antigens that are encoded in the HLA-D region and that stimulate secondary allogeneic proliferative and cytotoxic responses, *J. Exp. Med.*, 152, 565, 1980.
12. **Termijtelen, A., Bradley, B. A., and van Rood, J. J.,** A new determinant, defined by PLT, coded for in the HLA region and apparently independent of the HLA-D and DR loci, *Tissue Antigens*, 15, 267, 1980.
13. **Termijtelen, A. and van Rood, J. J.,** The role in primary MLC of the non HLA-D/DR determinant PL3A, *Tissue Antigens*, 17, 57, 1981.
14. **Andersen, L. C. and Hayry, P.,** Specific priming of mouse thymus-dependent lymphocytes to allogeneic cells in vitro, *Eur. J. Immunol.*, 3, 595, 1973.

15. **Fradelizi, J. and Dausset, J.,** Mixed lymphocyte reactivity of human lymphocytes primed in vitro. I. Secondary response to allogeneic lymphocytes, *Eur. J. Immunol.*, 5, 295, 1975.
16. **Sheehy, M. J., Sondel, P. M., Bach, M. L., Wank, R., and Bach, F. H.,** HL-A LD (lymphocyte defined) typing: a rapid assay with primed lymphocytes, *Science*, 188, 1308, 1975.
17. **Sasportes, M., Fradelizi, D., Nunez-Roldan, A., Wollman, E., Giannopoulos, Z., and Dausset, J.,** Analysis of stimulating products involved in primary and secondary allogenic proliferation in man. I. Preponderant role of the Ia-like DR (Ly-Li) antigens as stimulating products in secondary allogenic response in man, *Immunogenetics*, 6, 29, 1978.
18. **Jones, E. A., Goodfellow, P. N., Bodmer, J. G., and Bodmer, W. F.,** Serological identification of HL-A-linked human "Ia-type" antigens, *Nature (London)*, 256, 650, 1975.
19. **Mann, D. L., Abelson, L., Henkart, P., Harris, S. D., and Amos, D. B.,** Specific human B lymphocyte alloantigens linked to HL-A, *Proc. Soc. Natl. Acad. Sci. USA*, 72, 5103, 1975.
20. **Winchester, R. J., Fu, S. M., Wernet, P., Kunkel, H. G., Dupont, B., and Jersild, C.,** Recognition by pregnancy serums of non-HLA alloantigens selectively expressed on B lymphocytes, *J. Exp. Med.*, 141, 924, 1975.
21. **Bernoco, D., Bernoco, M., Ceppellini, R., Poulik, M. D., van Leeuwen, A., and van Rood, J. J.,** B cell antigens of the HLA system: A simple serotyping technique based on non-cytotoxic anti-beta-2-microgoblulin reagents, *Tissue Antigens*, 8, 253, 1976.
22. **Ting, A., Mickey, M. R., and Terasaki, P. J.,** B-lymphocyte alloantigens in Caucasians, *J. Exp. Med.*, 143, 981, 1976.
23. **Van Leeuwen, A., Schuit, H. R. E. and van Rood, J. J.** Typing for MLC. II. The selection of non-stimulator cells by MLC inhibition using SD identical stimulator cells (MISIS) and fluorescent antibody studies, *Transplant. Proc.*, 5, 1539, 1973.
24. **Bodmer, J. G., Pickbourne, P., and Richards, S.,** Ia serology, in *Histocompatibility Testing 1977*, Bodmer, W. F., Batchelor, J. R., Bodmer, J. G., Festenstein, H., and Morris, P. J., Eds., Munksgaard, Copenhagen, 1978, 35.
25. **van Rood, J. J., van Leeuwen, A., and Ploem, J. S.,** Simultaneous detection of two cell populations by two-colour fluorescence and application to the recognition of B-cell determinants, *Nature (London)*, 262, 795, 1976.
26. **Dausset, J. and Colombani, J., Eds.,** *Histocompatibility Testing 1972*, Munksgaard, Copenhagen, 1973.
27. **Sasazuki, T., McMichael, A., Payne, R., and McDevitt, H. O.,** HLA-D antigens in the Japanese population, *Tissue Antigens*, 9, 267, 1977.
28. **Stastny, P., Colbaugh, P., Heinle, A., Beathard, G. A., Lordon, R. E., Moraes, X., Jr., Oleinick, S. R., and Radvamy, R.,** Relation of human Ia antigens with other HLA products and functions in several American populations, in *Histocompatibility Testing 1977*, Bodmer, W. F., Batchelor, J. R., Bodmer, J. G., Festenstein, H., and Morris, P. J., Eds., Munksgaard, Copenhagen, 1978, 559.
29. **Troup, G. M., Jameson, J., Thomsen, M., Svejgaard, A., and Walford, R. L.,** Studies of HLA alloantigens of the Navajo Indians of North America. I. Variance of association between HLA-DRW (WIA) and HLADW specificities, *Tissue Antigens*, 12, 44, 1978.
30. **Layrisse, Z., Simoney, N., Park, M. S., and Terasaki, P. I.,** HLA-D and DRw determinants in an American indigenous isolate, *Transplant. Proc.*, 11, 1788, 1979.
31. **Reinsmoen, N. L., Sasazuki, T., Kaneoka, H., Ohta, N., Noreen, H. J., Greenberg, L. J., and Kersey, J. H.,** Two distinct HLAD specificities (DH0 and Dw2) in linkage with HLA-DRw2 as defined in White and Japanese populations, *Transplant. Proc.*, 10, 789, 1978.
32. **Vaidya, S., Colbaugh, P., and Stastny, P.,** Predictive value of HLA-DR matching for non-reactivity in mixed lymphocyte culture (MLC) depends on matching for HLA-D, *Transplant. Proc.*, 11, 1779, 1979.
33. **Bonnard, G. D., Chappuis, M., Glauser, A., Mempel, W., Baumann, P., Grosse-Wilde, H., and Albert, E. D.,** SD vs. LD antigens as targets for lymphocyte-mediated cytotoxicity: study of a family presenting a recombination event within the MHR, *Transplant. Proc.*, 5, 1679, 1973.
34. **Eijsvoogel, V. P., duBois, R., Melief, C. J. M., Zeylemaker, W. P., Raat-Koning, L., and de Groot-Kooy, L.,** Lymphocyte activation and destruction in vitro in relation to MLC and HLA, *Transplant. Proc.*, 5, 1301, 1973.
35. **Grunnet, N., Kristensen, T., Jorgensen, F., and Kissmeyer-Nielsen, F.,** Cell mediated lympholysis in man. The impact of SD determinants on *in vitro* induced lympholysis, *Tissue Antigens*, 4, 218, 1974.
36. **Geha, R. S., Malakian, A., Geha, O., and Yunis, E.,** Genetics of cell-mediated lympholysis in man, *J. Immunol.*, 118, 1286, 1977.
37. **Klein, J.,** Genetics of cell-mediated lymphocytotoxicity in the mouse, *Springer Semin. Immunopathol.*, 1, 31, 1978.
38. **Feighery C. and Stastny, P.,** HLA-D region-associated determinants serve as targets for human cell-mediated lysis, *J. Exp. Med.*, 149, 485, 1979.

39. **Feighery, C. F. and Stastny, P.**, Cell-mediated cytotoxicity against HLA-D region products expressed in monocytes and B lymphocytes. I. Blocking by unlabelled cells and inhibition by antisera, *Immunogenetics*, 10, 31, 1980.
40. **Feighery, C. F. and Stastny, P.**, Cell-mediated cytotoxicity against HLA-D region products expressed in monocytes and B lymphocytes. II. HLA-A or B-antigen compatibility is not required for HLA-D-region directed cytotoxicity, *Immunogenetics*, 10, 39, 1980.
41. **Ball, E. J., Feighery, C. F., and Stastny, P.**, Cell-mediated cytotoxicity against HLA-D region products expressed in monocytes and B lymphocytes. II. Inhibition by monoclonal antibodies and study of an HLA-B/DR recombinant, *Human Immunology*, in press.
42. **Ball, E. J., Okada, J., and Stastny, P.**, Characterization of cytotoxic lymphocytes against HLA-A, B, C and HLA-D/DR antigens, in press.
43. **Hartzman, R. J., Pappas, F., Romano, P. J., Johnson, A. H., Ward, F. E., and Amos, D. B.**, Dissociation of HLA-D and HLA-DR using primed LD typing, *Transplant. Proc.*, 10, 809, 1978.
44. **Shaw, S., Pollack, M. S., Payne, S. M., and Johnson, A. H.**, HLA-linked B cell alloantigens of a new segregant series: population and family studies of the SB antigens, *Human Immunol.*, 1, 177, 1980.
45. **Mann, D. L., Abelson, L., Harris, S., and Amos, D. B.**, Second genetic locus in the HLA region for human B-cell alloantigens, *Nature (London)*, 259, 145, 1976.
46. **Tosi, R., Tanigaki, N., Centis, D., Ferrara, G. B., and Pressman, D.**, Immunological dissection of human Ia molecules, *J. Exp. Med.*, 148, 1592, 1978.
47. **Terasaki, P. I., Ed.** *Histocompatibility Testing 1980*, UCLA Tissue Typing Laboratory, Los Angeles, Calif., 1980.
48. **Duquesnoy, R. J., Marrari, M., and Annen, K.**, Identification of an HLA-DR-associated system of B-cell alloantigens, *Transplantation Proc.*, 11, 1757, 1979.
49. **Lampson, L. A., and Levy, R.**, Two populations of Ia-like molecules on a human B cell line, *J. Immunol.*, 125, 293, 1980.
50. **Quaranta, V., Pellegrino, M. A., and Ferrone, S.**, Serologic and immunochemical characterization of the specificity of four monoclonal antibodies to distinct antigenic determinants expressed on subpopulations of human Ia-like antigens, *J. Immunol.*, 126, 548, 1981.
51. **Nunez, G., McPhaul, J., and Stastny, P.**, Characterization of B cell and monocyte antibodies in recipient sera and in eluates from rejected kidney allografts, *Transplant. Proc.*, in press.
52. **Colombani, J., Colombani, M., Dastot, H., Reboul, M., and Degos, L.**, Detection of human B lymphocyte alloantigens by complement fixation, *Transplantation*, 24, 230, 1977.
53. **Evans, R. L., Faldetta, R. J., Humphreys, R. E., Pratt, D. M., Yunis, E. J., and Schlossman, S. F.**, Peripheral human T cells sensitized in mixed leukocyte culture synthesize and express Ia-like antigens, *J. Exp. Med.*, 148, 1440, 1978.
54. **Yu, D. T. Y., Winchester, R. J., Fu, S. M., Gibofsky, A., Ko, H. S., and Kunkel, H. G.**, Peripheral blood Ia-positive T cell. Increases in certain diseases and after immunization, *J. Exp. Med.*, 151, 91, 1980.
55. **Gazit, E., Terhorst, C., and Yunis, E. J.**, The human 'T' genetic region of the HLA linkage group is a polymorphism detected on lectin-activated lymphocytes, *Nature (London)*, 284, 275, 1980.
56. **van Leeuwen, A., Festenstein, H., and van Rood, J. J.**, Human alloimmune sera against T cell subsets. Detection and influence on pokeweed mitogen-stimulated Ig production in vitro, *J. Exp. Med.*, 152, 235s, 1980.
57. **Okada, J. and Stastny, P.**, Alloantibody against a subset of human T cells, manuscript in preparation.
58. **Reinsmoen, N. L., Noreen, H. J., Friend, P. S., Giblett, E. R., Greenberg, L. J., and Kersey, J. H.**, Anomalous mixed lymphocyte culture reactivity between HLA-A, B, C, DR identical siblings, *Tissue Antigens*, 13, 19, 1979.
59. **Sachs, J. A., Jaraquemada, D., and Festenstein, H.**, Intra HLA-D region recombinant maps HLA-DR between HLA-B and HLA-D, *Tissue Antigens*, 17, 1981.
60. **Suciu-Foca, N., Godfrey, M., Rohowsky, C., Khan, R., Susinno, E., and Hardy, M.**, HLA-D-DR relationships. V. A crossover between HLA-D and DR, in *Histocompatibility Testing 1980*, Terasaki, P. I., Ed., UCLA Tissue Typing Laboratory, Los Angeles, Calif. 1980, 881.
61. **Jonker, M. and Balner, H.**, A review: current knowledge of the D/DR region of the major histocompatibility complex of Rhesus monkeys and chimpanzees, *Human Immunol.*, 1, 305, 1980.
62. **Stastny, P.**, Endothelial-monocyte antigens, *Transplant. Proc.*, 12, 32, 1980.

Chapter 7

Ia ANTIGENS OF RHESUS MONKEYS*

H. Balner

TABLE OF CONTENTS

*This work was carried out under contract no. 198-71-1 BION of the CEC's Radioprotection Program (paper no. 1769) and by support of the ZWO-FUNGO Organization.

I. INTRODUCTION

When asked to make a contribution with a chapter on the Ia antigens of nonhuman primates, I gladly accepted, however, with the request that the "deadline" for submission of manuscripts be postponed by about 6 months. I made that request because, in early 1981, it was particularly difficult for me to review the Ia antigens of rhesus monkeys and chimpanzees, the two primate species investigated. For *rhesus*, we had accumulated a wealth of empirical data on the DR locus antigens and several other Ia-like antigens, but there were still troublesome uncertainties regarding the nature and genetic control of some of those antigens. A new analysis using many new reagents and some new techniques had just begun in the spring of 1981 and several large new "harems" had become available for further testing. Hence, given an extra 6 months, we would have had the possibility to find the answers to at least some of the burning questions. With regard to *chimpanzees*, several D locus antigens had been only recently identified[1] and by mid-1981 an analysis of the available alloantisera reacting with B lymphocytes was under way. Again, more time would have been required to permit a reliable identification, characterization, and genetic analysis of the chimpanzee's DR antigens and possibly also of Ia antigens not controlled by the DR locus. Unfortunately, postponement of the editors "deadline" was impossible. Hence, this contribution will review only the DR and other Ia antigens of rhesus monkeys as we know them in early 1981, with emphasis on their serology, genetics, and possible significance in transplantation immunology. Information on Ia antigens of chimpanzees will be published elsewhere, in due course.

I shall refrain for obvious reasons from philosophizing about the role of Ia antigens in immunobiology.* First, rhesus monkeys and chimpanzees are hardly the ideal species for fundamental work in that intricate field. Secondly, this volume is certain to contain contributions from many investigators who have meticulously studied the genetic control, biochemistry, and biological significance of Ia antigens in species which lend themselves far better than primates to such sophisticated work. We have followed that work with great interest and the more we learn about Ia antigens of rodents, the more convinced we are that what holds true for Ia antigens of the mouse, the rat, or the guinea pig, also holds true for Ia antigens of the primates, including, man. Therefore, we impatiently await the final word of the experts on the chemical fine structure and genetic control of Ia antigens, their role in cell-to-cell interactions, and on the vital question of whether "the hybrid Ia antigens generated by the interaction of two polypeptide chains (in the mouse coded by genes in the I-A and the I-E subregions of H-2) are indeed essential for immune recognition and are thus the true immune response or Ir gene products."[2]

II. ANIMALS AND METHODS EMPLOYED

A. Animals

The *unrelated* rhesus monkeys used were all imported from India in a number of shipments between 1965 and 1975. Kinship among animals of each shipment is unlikely but cannot be excluded. The *related* monkeys were all raised in the Primate Center TNO, employing a harem-type breeding system with approximately ten females per breeding male. Hence, all offspring of a particular male are sibs or half-sibs; all will carry the paternal RhLA haplotypes designated a or b and the maternal RhLA haplo-

*Apart from reviewing the clinically relevant data on the influence of host/donor matching for DR antigens on the fate of various types of allografts in rhesus monkeys (see Section VI of this chapter).

types c or d from the same or from different mothers. A few backcrosses of daughters to fathers (and their offspring) were also involved in the family studies reviewed.

B. Typing

1. Serological Techniques

"SD-typing" for the conventional A and B locus antigens was performed with a modified Kissmeyer-Nielsen one-stage microcytotoxicity assay.[3] The methods of producing, selecting, and screening optimal antisera have been described in detail.[4,5] Typing for DR and other B cell-specific or Ia-like antigens was performed by application of a two-stage lymphocytotoxicity technique using B cell enriched lymphocyte suspensions.[6] The reagents were produced by immunizing rhesus monkeys with skin grafts and subsequent blood injections from a single donor sharing all four A/B antigens and (at a later stage) from A/B identical donors sharing also one or two DR antigens. The screening, selection, and further characterization of the reagents have been fully described.[7–12]

2. Cellular Techniques

A modification of the microtechnique for mixed lymphocyte cultures described by Balner and Toth [13] was used. Heparinized blood was diluted 1:1 in medium (Hanks' solution supplemented with 100,000 U/ℓ penicillin, 0.1 g/ℓ streptomycin and buffered with 15 m*M* Hepes). The lymphocytes were separated by gradient and suspended in culture medium. Stimulator cells were exposed to 25 Gy of irradiation. The one-way tests were performed in triplicate using 10^5 responder and 10^5 stimulator cells. After 96 hr of incubation at 37° C, 0.08 μCi of ^{14}C-TdR was added; cells were harvested 24 hr later and uptake of ^{14}C-TdR was determined by liquid scintillation counting.

3. The Use of Homozygous Typing Cells to Identify D Locus Antigens

This method has been extensively described.[14,15] It is based on the principle that the D locus antigens have an overriding influence on MLC responsiveness. Thus, lymphocytes from an individual which is heterozygous for the D locus antigens will show a negative or low MLC response when tested unidirectionally against a cell sample homozygous for either of the two D antigens in question. Reversely, the homozygous "typing cell" will react normally in MLC to the stimulating antigen which it does not share with the heterozygous cell sample. Hence, a TC-response is asymmetric: negative in one direction (TC as stimulator) and positive in the other (TC as responder).

4. Evaluation of MLC Data

The MLC results were expressed as stabilized relative responses or SRR.[16] Briefly, each responder cells'response, without correction for autologous values, was calculated relative to the 75th percentile of all cpm values (RR). Next, the responses were calculated relative to the 75th percentile of the RR values of each stimulator cell (SRR). MLC nonreactivity was based on the outcome of MLC tests between RhLA identical siblings (below 40%, with a mean of 20%). An SRR value of 50% or less was considered as a typing cell response, one of 50 to 60% as a questionable TC response, and a higher one as a positive MLC. For each particular MLC combination, the median value of the SRRs of repeat experiments was used.

C. Genetic Analysis

The genetic analyses were carried out in several steps, using methods established during the past 20 years for analyzing MHC products of outbred species. The first step is usually a cluster analysis of the reactivity patterns of available antisera or other reagents (e.g., typing cells) against the relevant cells of a large number of unrelated individuals.[17] Once tentative "antigens" have been assigned, their relationship is investigated by determining associations in one or more populations of those unrelated animals. Absorptions of serological reagents may be required to improve the definition of the antigens. Subsequently, certain characteristics of the antigens are established and their genetic relationship is investigated by conventional 2 × 2 chi-square tests for "independence" and for "allelism"; potential alleles of a locus are tested for their fit to Hardy-Weinberg equilibrium.[10,18] The gene frequencies required for the test are usually determined by maximum likelihood estimates from the phenotype data.

The second step is a segregation analysis of the identifiable antigens in families. In our case (rhesus or chimpanzee), the analysis is simplified by the availability of multiple harems, each consisting of numerous families. Family data will provide reliable information on the genetic control of the antigens by separate loci, on linkage among those loci, on linkage disequilibria between antigens of the linked loci,* and on the genetic mapping of the genes or loci within a genetic region. After linkage of different genes or loci in a region (such as the MHC) has been established, a "recombinant" offspring will usually provide information on the mapping position of the genes and/or loci vis-à-vis each other.

D. Transplantation

1. Kidney Transplantation

Kidney transplantation was performed as described in detail elsewhere.[20] The donor renal vessels were anastomosed end to side with the recipient's aorta and vena cava; recipients were bilaterally nephrectomized immediately after transplantation. Death attributable to graft failure was considered as the endpoint of graft survival. Histopathological examinations were always possible for determination of the cause of renal failure.

2. Skin Grafting

This technique has also been fully described.[21] In the evaluation of the effect of D/DR matching on graft survival, only data for skin grafts obtained from a single unrelated donor were considered. Briefly, 2 grafts of about 2 cm in diameter are taken from the donor's abdomen and after removal of subcutaneous tissue, transplanted to the recipient's back. Grafts are fixed with sutures and a special dressing is used to immobilize the area and exert pressure on the grafts. From day 5 on, grafts are inspected on alternate days, from day 10 on, daily. Major hemmorrhages and crust formation over most of the graft area was taken as the endpoint of graft survival.

3. Bone Marrow Transplantation

The rather intricate procedure for transplantation of allogeneic bone marrow (or stem cells) into irradiated monkeys has also been described at length.[22] Briefly, young rhesus monkeys (2 to 3 years) are conditioned with 850 cGy total body X-irradiation (TBI). More recently, better "takes" of the graft were obtained when 2 × 700 cGy X-ray TBI were given with a 3-day interval between doses. Allogeneic stem cell concentrates are given within 24 hr following the last radiation (1 to 5 × 10^7 cells/kg). After transplantation of marrow or stem cells, measures are taken to protect the animals against infection (isolation, antibiotics) and hemorrhages (platelet transfusions). They are ob-

*Linkage disequilibria between products of various linked loci (in our case, loci controlled by the major histocompatibility complex) can also be derived from population data.[19]

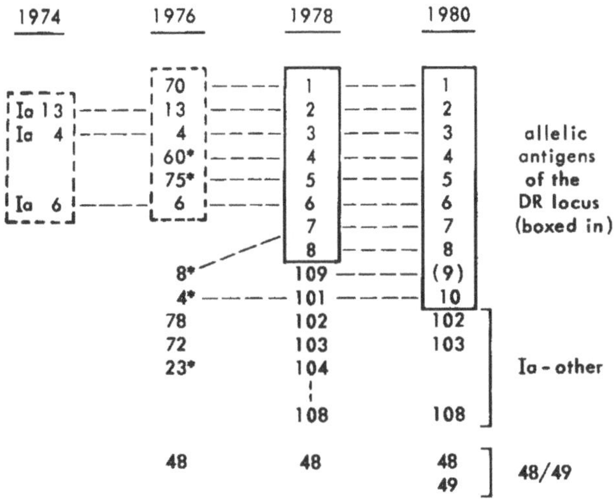

FIGURE 1. DR and other B cell antigens of rhesus monkeys. Chronology of identification and changes of nomenclature. Identical DR antigens are connected by interruped lines. For further details see text and the cited reference.

served for take of the graft, development of fatal or nonfatal Graft-vs.-Host Disease (GvHD), and the eventual recovery of the immune system in the long-surviving chimeras.

III. SEROLOGICALLY DEFINED B CELL ANTIGENS

Figure 1 summarizes the "history" of the identification and nomenclature of the serologically defined B cell antigens of rhesus monkeys. Briefly, the first Ia-like antigens were identified in 1974. Unlike in man, the antisera were produced by intentional immunization of monkeys with tissues from donors identical for the four A and B locus antigens.* The first antigens identifiable with such sera were designated as 4,6, and 13; they were proposed to be the primates' analogues of the murine Ia antigens (serology, genetics, tissue distribution, co-capping studies, etc.).[7,8] Subsequent production and screening of many similar reagents permitted the identification of a dozen Ia-like antigens by 1976 (Figure 1, column 2). Population and family studies then indicated that at least 6 of the specificites were controlled by an RhLA-linked segregant series which was called the Ia_1 locus.[10] several of the other identifiable B-cell antigens were viewed as candidate members of a second RhLA-linked series of Ia antigens. By 1978 (Figure 1: column 3), the availability of many more reagents and another genetic analysis suggested 8 alleles for the major Ia locus and 1 or more RhLA-linked loci controlling some of the "other" B cell antigens.[11] A close association was found between the Ia_1 antigens and D locus antigens defined by cellular methods.[24] Therefore, the name of the major Ia locus was changed from Ia_1 to DR (D-Related), in accord with the nomenclature proposed for the principal locus of human Ia antigens during the Oxford workshop of 1977.[25] With regard to the "other" Ia antigens of the monkey,

*At a later stage, this approach of intentional immunization was also used by investigators working on Ia antigens of man.[23]

specificity 8* now appeared to be another allele of the DR locus and was renamed DR 7. The numerical symbols for several other B cell antigens were also changed, but their genetic control remained undefined. Results of more recent studies, however, provided indications for the probable genetic control of also those other Ia antigens (Figure 1, column 4). In this chapter, some of the relevant figures and tables of a population study carried out in 1979,[12] and of a family study performed with the same reagents in 1980,[26] will be presented.

At the time of this writing, yet another genetic analysis of the Ia antigens of rhesus monkeys is in progress. Of the many new reagents which are being used, some were raised by cross-immunization of RhLA-identical siblings and detect antigens not controlled by the RhLA region. Results of that study, including updated information on composition and mapping positions of the DR locus and the "other" Ia locus or loci (vis-à-vis previously established genes and loci of RhLA) are to be published soon.

A. The DR Locus

Data based on a *population study* performed in 1979 are shown in Figure 2 and Table 1. The figure graphically depicts the distribution of the 8 DR antigens in a population of 160 unrelated rhesus monkeys, and serves to demonstrate serological relationships among several groups of antisera and antigens. Thus, DR5, 6, 7, and 8 constitute one such "cross-reactive" group; the boxed-in areas indicate that the sera defining DR5 also react with cells carrying DR8, 6, 7. Likewise, anti-DR6 sera also react with cells carrying DR7. A second group of cross-reacting antigens is made up of DR3 and 4; since DR3 is "included" in DR4, antigen DR4 cannot be identified in the presence of DR3. A third group is probably formed by DR1 and 2, since the majority of sera raised against DR1 and/or DR2 contain antibodies directed against both antigens; however, in this case, some reagents which permit the separate distinction of DR1 and DR2 are available. Therefore, in Figure 2, no "blind spots" (boxed-in areas) are given for antigens DR1 and DR2.

The long or "broad" reagents described above (anti-5 sera, anti-4 sera, and sera defining DR1 + 2) are reminiscent of similarly broad reagents defining the antigens of the MB or MT loci of the human HLA system.[27,28] In the monkey also, the antigens defined by those broad reagents seem to behave as alleles of a genetic system, although recombinants between genes of the DR-locus and that hypothetical MB/MT-like locus of the monkey have not been found. Yet, the very similar organization of the human HLA region and the analogous RhLA region of rhesus makes it tempting to speculate that a separate MB/MT locus also exists in rhesus monkeys, although proof for that assumption is missing. Incidentally, the other B cell antigens controlled by the RhLA region (102, 103, 108, 48, 49) are not likely candidates as alleles of the hypothetical MB/MT locus of the monkey; they have low gene frequencies and do not show the typical inclusions of two or more DR antigens as reported for the antigens of the human MB/MT locus.

Table 1 shows the phenotype distribution for DR antigens of the animals depicted in Figure 2 and compares the combinations observed with those expected on the basis of estimated gene frequencies (maximum likelihood method; from Roger et al.[12] The fit to Hardy-Weinberg equilibrium was reasonable. Further, antigens 4* (later called 101) and at least part of specificity 109 (see Figure 1) were considered as candidates for the DR series, since both were highly associated with "blanks" of the DR locus[11,12] (see below).

A *family study* using reagents available for the population study of 1979 provided further information regarding the genetics of the DR antigens and the "other" B cell

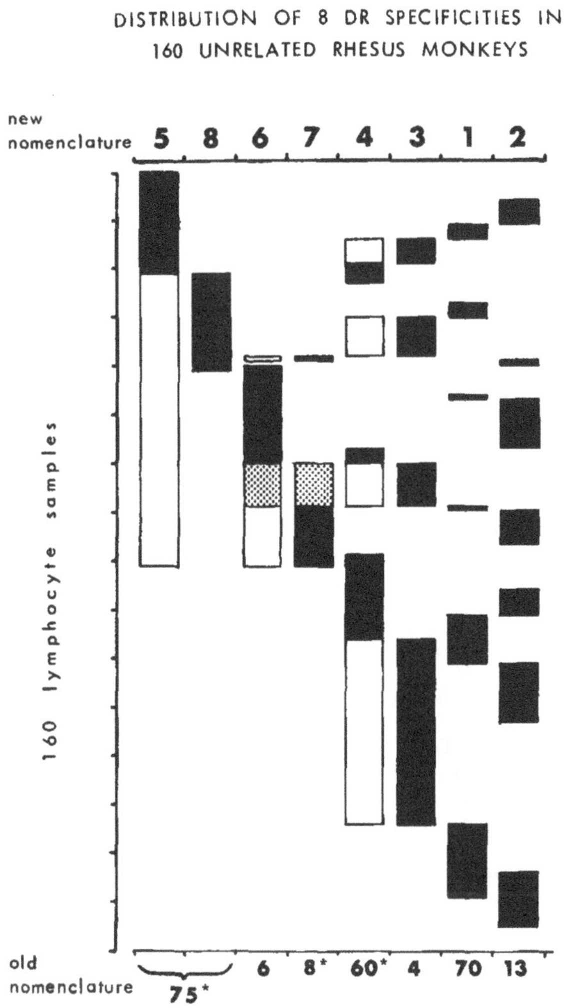

FIGURE 2. Distribution of eight DR specificities in 160 unrelated rhesus monkeys. A black horizontal bar indicates that a lymphocyte sample reacted positively with the majority of antisera defining a specificity (four to ten serum batches from two to five different monkeys). Among the cross-reacting DR specificites, antigen 4 cannot be defined in the presence of 3, antigen 5 in the presence of 6, 7, or 8 (anti-5 sera are the "broadest"); antigen 6 in the presence of 7 and antigen 7 in the presence of 3 (the "double-blind" sector of antigens 6/7 is shaded). (From Roger, J. H., van Vreeswijk, W., and Balner, H., *J. Immunogenet.*, 7,333, 1980. With permission.)

antigens. Tables 2 and 3 present a selection of family data obtained during 1980.[26] When 8 of the largest "harems" (consisting of 8 males, 88 females, and about 200 of their offspring) had been analyzed, a number of tentative conclusions could be made.

The DR series as conceived on the basis of previous population studies is basically correct. However, DR6 and 8 show a peculiar relationship: the 2 antigens occurred together on 3 of a total of 162 haplotypes analyzed; 1 such case is shown in Table 2,

Table 1
PHENOTYPE DISTRIBUTION ON CELLS OF 160 UNRELATED RHESUS MONKEYS OF 8 PROPOSED SPECIFICITIES CONTROLLED BY THE DR LOCUS

Gene freq.	First/second specificity	Observed	Expected
.112	DR 1/DR 2	5	5.99
	1/ 3	5	7.47
	1/ 4	5	3.54
	1/ 5	3	2.89
	1/ 6	1	3.02
	1/ 7	1	1.86
	1/ 8	3	2.30
	1/Blank	10	6.86
.167	DR 2/DR 3	12	11.07
	2/ 4	5	5.24
	2/ 5	5	4.29
	2/ 6	10	4.48
	2/ 7	7	2.76
	2/ 8	1	3.42
	2/Blank	6	11.61
.208	DR 3/DR 5	5	5.35
	3/ 6 or 7	9	9.04
	3/ 8	8	4.26
	3/ 4 or Blank	21	22.39
.098	DR 4/DR 5	2	2.53
	4/ 6	3	2.65
	4/ 7	2	1.63
	4/ 8	2	2.02
	4/Blank	5	5.78
.080	DR 5/Blank	6	4.50
.084	DR 6/DR 8	1	1.72
	6/ 5 or Blank	5	6.91
.052	DR 7/DR 8	1	1.06
	7/ 5 or 6 or Blank	2	5.39
.064	DR 8/DR 5 or Blank	4	5.07
.135	Blank/Blank	5	2.90
		160	160.00

Note: $X^2_{22} = 31.88$ ($0.10 > P > .05$) for goodness of fit to Hardy Weinberg. The reasons for "pooling" certain combinations of specificities are described in Roger et. al.[12] Two animals in the total sample of 162 appeared to express three of these DR specificities (DR2/7/8 and DR2/3/6) and were excluded from this table.

From Roger, J. H., van Vreeswijk, W., and Balner, H., *J. Immunogenet.*, 7, 33, 1980. With permission.

in the back-cross family ♂600 × ♀S-600. This relationship between antigens 6 and 8 is particularly interesting in view of a similar "association" observed in cellular typing for D antigens (see below). Possible explanations for the relationship between D/DR 6 and 8 have been published.[26,29]

Antigens 101 and 109 are still interesting candidates for the DR series. When occurring, specificity 101 was associated with a blank of DR on the haplotypes (e.g., harem 2774 in Table 2). Specificity 109 is not as well defined serologically and presented "problems" on several occasions (examples are shown in Table 3). However, there is also the observation that the group of "typing cells" defining antigen D9 all react positively with anti-109 sera (see later, Figure 4), and the finding of a positive association of antigen D9 with 109 in a population of rhesus monkeys.[29] Such independent observations strongly suggest that at least some of the antibodies currently defining 109 identify a DR locus antigen which, in anticipation of the availability of more restricted reagents, has been provisionally named DR9 (4th column of Figure 1; 9 in parentheses).

There were several interesting *linkage disequilibria* among DR antigens and various other classes of antigens. Table 4 shows linkage disequilibria for the combinations where the values are greater than 0.01 and significantly greater than zero ($p < 0.05$). Detailed discussion has been published elsewhere.[12,26] Suffice it to mention here that several linkage disequilibria found previously in population studies were confirmed in the haplotypes in this investigation; this includes a significant delta between antigens B23 and DR6.[11]

Of particular interest were disequilibria involving DR antigens and three other classes of RhLA-controlled markers. It was found that 5of the 6 haplotypes with the A11-B9 combination (a well-known linkage disequilibrium among class I antigens of RhLA observed in several previous studies[3]) also carried 108 ("Ia-other"), while 4 of those 5 haplotypes also carried antigen DR2. Further, of the 6 haplotypes which carried A11, B19, and antigen 48, 5 had DR3 as well! A provisional mathematical analysis of these 2 special categories of haplotypes revealed that the combination 108/DR2/A11/B19 can be explained in terms of linkage disequilibria involving only pairs of alleles (first-order interactions). However, the combination DR3/B19/A11/48 suggests that there is also a second-order interaction, so that this particular haplotypes may have an increased frequency also in the population at large.[31] It is not unreasonable to speculate that such "super-haplotypes" will have biological significance.

B. Characteristics of DR Antigens

Certain characteristics of DR antigens and the antisera defining them have been published. It was shown by co-capping studies that Class II or Ia antigens of rhesus monkeys are distinct from Class I or A and B locus antigens, and that Ia antigens are not associated with β 2 microglobulin.[8] In addition, absorption studies on Ia antisera had revealed that Ia or DR antigens have a tissue distribution similar to that of the murine Ia antigens.[7] Likewise, anti-DR sera of the monkey showed specific inhibition of the stimulator function of lymphocytes in MLC[9] as do analogous Ia antibodies in mice[32] and man.[6] Thus, various lines of evidence strongly suggest that the DR antigens of the rhesus monkey are the analogues of the DR antigens of man and the Ia antigens of the mouse.

The biochemical analysis of the Ia antigens of the monkey is still incomplete. There is solid evidence that the DR antigens of rhesus monkeys are chemically distinct from the A or B locus antigens of that species. The latter have been defined biochemically[33] and appear to have the same biochemical characteristics as class I or SD antigens of other species. Further, it can be stated with confidence that DR antigens of the monkey

Table 2
GENOTYPING OF THE RHESUS MONKEY FAMILIES BELONGING TO THE "HAREMS" OF ♂600 AND ♂2774

Parents and offspring	AB geno-types	Genotyping for antigens other than AB		
		DR	Ia-other	48–49
♂600	ab	3/(4)	102/102	—/—
♀597	cd	—/2	109/108	—/—
AB	bc	4/—	102/109	—/—
BF	bc	4/—	102/109	—/—
DW	ac	3/—	102/109	—/—
HO	ad	3/2	102/108	—/—
IW	bd	4/2	102/108	—/—
♀669	cd	8/4	—/103	—/—
AC	ad	3/(4)	102/103	—/—
BI	ad	3/(4)	102/103	—/—
CR	ad	3/(4)	102/103	—/—
EL	bc	4/8	102/—	—/—
FS	bc	4/8	102/—	—/—
♀Y-381	cd	8/3	102/?	—/—
HB	ac	3/8	102/102	—/—
MY	ac	3/8	102/102	—/—
FA	ad	3/3	102/?	—/—

Parents and offspring	AB geno-types	Genotyping for antigens other than AB		
		DR	Ia other	48–49
♂2774	ab	3/—	—/101	48/—
♀2837	cd	4/—	—/109	—/—
OF	ac	3/(4)	—/—	48/—
WE	ac	3/(4)	—/—	48/—
→RX	a/bc	3/(4)	—/—	—/—
JF	bc	—/4	101/—	—/—
ME	bd	—/—	101/109	—/—
♀2797	cd	7/2	—/108	49/—
JV	ac	3/7	—/—	48/49
QU	ad	3/2	(—)/108	48/—
NQ	bc	—/7	101/—	—/49
♀2829	cd	1/2	—/108	—/—
SG	ac	3/1	—/—	48/—
PD	ad	3/2	(—)/108	48/—
VA	bc	—/1	101/—	—/—

HS	ad	3/3	102/?	—/—
GC	ad	(4)/3	102/?	—/—
♀S-600	bd	4/6 + 8	102/?	—/—
EX	bd	4/6 + 8	102//?	—/—
LY	bd	4/6 + 8	102/?	—/—
JD	ab	3/(4)	102/102	—/—
♀2504	cd	(2/2)	(108/108)	(—/—)
IU	bc	4/2	102/108	—/—
VM	bc	4/2	102/108	—/—
SK	bd	4/2	102/108	—/—
LJ	ad	3/2	102/108	—/—
♀2596	cd	6/2	103/108	—/—
PT	bc	(4/6)	(102/103)	(—/—)
TL	bc	4/6	102/103	—/—
VT	bd	4/2	102/108	—°—
♀852	cd	—/—	—/—	—/—
IJ	bd	4/—	102/—	—/—
TW	ad	3/—	102/—	—/—
RS	ac	3/—	102/—	—/—

♀2801	cd	5/—	—/102	—/—
MK	bc	—/5	101/—	—/—
TI	bd	—/—	101/102	—/—
				—/—
♀2679	cd	3/2	—/—	—/48
WC	bc	—/2	101/—	—/—
♀2835	cd	5/4	—/103	—/49
QB	bd	—/4	101/103	—/49
♀2810	cd	3/1	—/?	48/?
JO	bc	—/1	101/—	—/48
♀2709	cd	1/2	—/108	—/—
OV	ac	3/1	—/—	48/—

Note: Symbols a and b stand for paternal, c and d for maternal RhLA haplotypes as defined by the antigens of the A and B loci. Data for genotyping are placed in brackets when deduced (animal not typed) or when the presence of an antigen could not be demonstrated seologically (e.g., DR4 in the presence of DR3). A "slash" separates the paternal (left) from the maternal (right) contributions. A query indicates that homozygosity cannot be excluded. Arrows indicate that the offspring in question is probably a recombinant. Further details are discussed in the text and in Balner et al.[26]

Reproduced from Balner, et al., *Tissue Antigens*, 17, 530, 1981. With permission.

Table 3
GENOTYPING OF THE RHESUS MONKEY FAMILIES BELONGING TO THE "HAREMS" OF ♂2315 AND ♂2775

Parents and offspring	AB genotypes	Genotyping for antigens other than AB: DR	Ia-other	48–49
♂2315	ab	2/3	108/—	49/—
♀2598	cd	—/3	—/—	—/48
JN	ac	2/—	108/—	49/—
PS	ac	2/—	108/—	49/—
XQ	ac	2/—	108/—	49/—
MX	ad	2/3	108/—	49/48
→RY	bd	2/3	108/—	—/48
♀1115	cd	—/3	102 + 109/109	—/—
OJ	ac	2/—	108/102 + 109	49/—

Parents and offspring	AB genotypes	Genotyping for antigens other than AB: DR	Ia other	48–49
♂2775	ab	8/1	—/102	—/—
♀2771	cd	8/3	109/102	—/—
QE	ac	8/8	—/109	—/—
UP	ad	8/3	—/102	—/—
SB	bc	1/8	102/109	—/—
♀2854	cd	1/3	—/—	—/—
PC	ac	8/1	—/—	—/—
JY	bc	1/1	102/—	—/—
SQ	bd	1/3	102/—	—/—

LO	bc	3/—	—/102 + 109	—/—
HY	bd	3/3	—/109	—/—
♀1644	cd	(4)/3	—/102	—/48
IL	ac	2/4	108/—	49/—
PJ	ac	2/4	108/—	49/—
→KP	bc	3/(4)	—/—	49/—
♀2601	cd	1/1	(—/—)	(49/—)
IS	bc	3/1	—/—	—/49
LU	bc	3/1	—/—	—/49
UF	bd	3/1	—/—	—/—
♀2470	cd	?/3	—/—	—/—
OW	bc	3/?	—/—	—/—
→TF	a/bd	2/3	108/—	—/—

♀2354	cd	5 or 8/3	?/103	—/—
SH	ad	8/3	—/103	—/—
WL	ad	8/3	—/103	—/—
OR	bd	1/3	102/103	—/—
♀2820	cd	4/2	108/108	—/—
SU	ad	8/2	—/108	—/—
WG	ad	8/2	—/108	—/—
KJ	bc	1/4	102/108	—/—
♀2822	cd	1/3	—/108	—/—
→PX	ac	8/3	102/108	—/—
TJ	bd	1/3	102/108	—/—
♀2818	cd	3/5 or 8	102	—/48
→MA	bc	8/3	102	—/—

Note: For explanation of signs and symbols, see note of Table 2.

[a]Families with one or two offspring are depicted only if recombinations are involved.
Reproduced from Balner et al. *Tissue Antigens,* 17, 530, 1981. With permission.

Table 4
LINKAGE DISEQUILBRIA OBSERVED IN RhLA HAPLOTYPES FOR CERTAIN PAIRS OF ANTIGENS

	RhLA-controlled antigens					Combinations			
Comparisons	A	B	DR	Ia-other	48–49	observed ++	expected ++	Δ	P
A with B	11	6				8	4.3	.023	.0039
	13	6				5	1.4	.022	.0043
	14	5				3	0.8	.014	.0351
A with DR	2		4			3	0.4	.016	.0047
	11		2			9	4.6	.027	.0195
	18		4			3	0.5	.015	.0089
	34		6			4	0.5	.021	.0004
B with DR		23	6			4	1.1	.018	.0120
		blank	3			4	1.4	.016	.0288
A with Ia-other	11			108		10	5.7	.026	.0289
B with Ia-other		9		108		12	5.9	.037	.0035
DR with Ia-other			2	108		21	6.9	.087	.0001
			4	103		5	1.0	.024	.0013
			8	102		5	1.4	.022	.0071
			blank	101		2	0.2	.011	.0131
			blank	109		6	1.9	.025	.0045
A with 48–49	11				48	13	2.6	.064	.0001
	17				49	8	0.6	.047	.0001
B with 48–49		19			48	6	2.6	.021	.0288
DR with 48–49			3		48	7	3.4	.023	.0284

Note: Antigen combinations showing a significant linkage disequilibrium (Δ) with "exact P" values derived from the one-sided Fisher test ($p < .05$); only Δ values > 0.01 are represented in the Table. The data are based on an analysis of 162 parental RhLA haplotypes.

From Balner, H., van Vreeswijk, H., and Roger, J. H., *Tissue Antigens,* 17, 530, 1981. With permission.

are not associated with β2m and that, as in the other species investigated, they consist of two polypeptide chains.[30] The molecular weights of those 2 chains are in the order of magnitude of 30,000 daltons, but exact figures for the monkey's analogues of the α and β chains of Ia molecules are as yet undetermined.

C. Ia-Like Antigens Not Defined by the DR Locus

As indicated, a number of Ia-like antigens identifiable on monkey lymphocytes are not controlled by the DR locus. The specificities known in early 1981 are depicted in Figure 1 and in Tables 2 through 4. Similar to DR antigens, such non-DR, Ia-like antigens are controlled by the RhLA region, are most readily demonstrable on B lymphocytes, and the antibodies to identify them are also raised by immunization of animals with tissue from a donor sharing the four A and B locus antigens, usually also compatible or identical for the DR antigens. In cases where absorptions of such reagents were done, the relevant antibodies could not be removed with platelets.

While some of the non-DR, Ia-like antigens subsequently proved to "fit" the DR series (see Figure 1 and the section on the DR locus), others have remained separate serological antities whose characteristics, genetics, and chemical structure are as yet poorly defined. Some of the "Ia-other" antigens accepted in 1978 had to be dropped from the list because of unsatisfactory reproducibility of their serological definition in consecutive studies. Thus, the provisional antigens 104, 105, 106, and 107[11] have been disregarded in subsequent studies.

The family study carried out in 1980 permitted a further provisional subdivision of the non-DR antigens into the group referred to as "Ia-other" (102, 103, and 108) and the category designated as 48–49.[26] Segregation data, including those for several recombinant offspring, suggest that both sets of antigens are controlled by the MHC but by different sections of the RhLA region (see below). A biochemical analysis of the non-DR Ia antigens is in progress; preliminary results suggest that 48 is biochemically distinct from the conventional Class I and Class II antigens.[30] Incidentally, at the time of this writing, a class of antigens analogous to 48 of the monkey is not known for mouse or man. For these reasons, and because of peculiar characteristics of at least some of the anti-48 sera (occasional inhibition of responder cells in MLC[9]), antigen 48 and its possible allele 49 are listed separately from DR and from the "Ia-other" antigens in the figures and tables of this chapter.

D. Genetic Mapping of Genes and Loci Defining Ia-Antigens in Rhesus Monkeys

Figure 3 presents the current (early 1981) concept of the genetic mapping positions of the genes and loci of RhLA discussed so far. (A description of the D locus and its relation to DR is given in the following sections). The model is based on family studies carried out between 1975 and 1980. The eight recombinant offspring which provided the data for genetic mapping are shown below the schematic model of the RhLA region in Figure 3. Information on several RhLA-linked immune response genes is also available;[36,37] those Ir genes are not included in the model of Figure 3 because their mapping positions are not accurately defined, although there is provisional evidence that the genes regulating the responses to the synthetic antigens GL and GA map "to the left" of the A/B region, in the vicinity of D/DR.

Detailed information on the recombinant offspring depicted in this figure can be found in the literature; for complement factor Bf (animals AX and DS), see Ziegler et al.,[35] for data on offspring AX and LD, see van Es and Balner;[34] for the other recombinants, see Table 2 and 3 of this chapter and Balner et al.[26]

In making a comparison between the genetic maps of the MHC of man and the rhesus monkey (Figure 3), we may conclude that both species have A and B loci with a recombination frequency of less than 1%. The polymorphism of the A and B loci is somewhat greater in man, but the A/B locus products are biochemically similar for the two species. No analogue for the human C locus has yet been found in the monkey. The region "to the left" of A and B contains the D/DR locus or loci and may be regarded as the I region of primates.[7] In the monkey and man, the recombination frequency between A/B and D/DR is slightly less than 1%. Whether D and DR antigens are identical (same gene?, same molecule?) is still not known for the monkey and for man.[8] What are considered here as Ia-other and 48/49 for the monkey (Figure 3) are, to our knowledge, not known for man. The separate MB or MT locus proposed for man[27,28] is not represented in Figure 3, since the evidence for an analogous locus in the monkey is still tenuous. Likewise, while there is evidence for one or more MLC loci other than D (but also controlled by the MHC region) in the rhesus monkey and in man, the mapping position of those loci is still too uncertain to be depicted in a model such as that of Figure 3. Direct evidence for the presence of MHC-linked Ir genes has been obtained for the monkey while such evidence for man is still indirect;

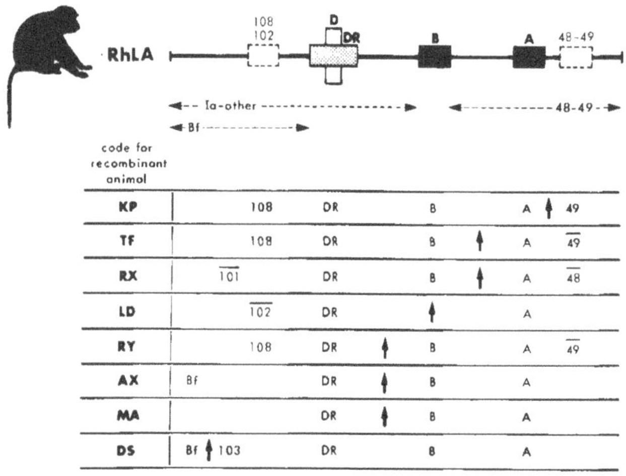

FIGURE 3. Graphic display of the proposed mapping position of RhLA-linked genes. Detailed data for the recombinant animals KP, TF, RX, RY, and MA are given in Balner et al.;[26] data for animals LD, AX, and DS in van Es and Balner[34] and Ziegler et al.[35] In the lower part of the figure, symbols for RhLA loci (or their products) are given only when relevant information is provided by the recombinant animal. Bold vertical arrows indicate the proposed site of recombination. A horizontal dash above the numerical designation of an antigen means that the antigen in question was not expressed, while the parent (from which the recombinant haplotype stems) did carry at least one antigen of the proposed series. In the graphic model of RhLA at the top of the figure, the mapping positions of the A, B, and D/DR loci are fairly certain; those of the two other postulated loci (102–108 and 48–49) are still uncertain. In view of the uncertainty about control of 48 and 49 by a single locus, the horizontal arrow indicating the position of those genes within RhLA is extended to the B locus. Distances between loci and antigens are not based on estimated recombination frequencies. The mapping positions of several RhLA-linked Ir genes[36] remain uncertain and are not depicted in this figure. (From Balner, H., van Vreeswyk, W., and Roger, J. H., *Tissue Antigens*, in press. With permission.)

for both species, the Ir genes are believed to be located in the vicinity of D/DR. Finally, with regard to complement factors, a locus for Bf is present in man and in the monkey in the region to the left of A/B. Complement factors C2 and C4 which are HLA-controlled in man have not yet been shown to be polymorphic in the rhesus monkey. Similarly, the enzyme GLO which is MHC-linked in man[39] is not polymorphic in the rhesus monkey.

IV. ANTIGENS DEFINED BY CELLULAR TECHNIQUES

A. Homozygous Typing Cells

The principle of defining the D locus antigens by means of D-homozygous typing cells (TC) has been explained in Section II of this chapter. Three types of TCs were used: lymphocytes obtained from (1) consanguinious animals carrying two identical RhLA haplotypes; these were offspring of matings among sibs, half-sibs, or parents and offspring; (2) animals from the breeding colony shown to be phenotypically homozygous for DR locus antigens; and (3) imported, feral animals carrying a single DR antigen or only a blank of that series, animals which are thus potentially homozygous for DR. Cells of types 2 and 3 were accepted as TCs only when showing consistent "asymmetrical" MLC reactions with cells from heterozygous animals.*

Table 5 lists the 38 TCs available in 1980 for the cellular definition of D locus antigens. The table shows (from left to right) their grouping according to D specificities and their numerical designations (1 to 38), their origin (*i*nbred, *g*enotyped, or *f*eral; see above) and their phenotypes or genotypes for the A-, B-, and DR-loci, as well as for the other B cell antigens (102, 103, and 108, 48/49). The group designation of typing cells (D1, D2, etc.) is clearly related to their phenotypes or genotypes for DR antigens. Further details are given elsewhere.[29,34,41]

Certain characteristics and the "homogeneity" of these 10 groups of TCs were investigated by performing MLC tests of the 38 cells against each other in matrix-type fashion.[29] MLCs among TCs of the same group were performed 5 to 7 times; among TCs of different groups 2 to 4 times. The left-hand side of Figure 4 shows the results. It can be seen that MLCs among cells carrying the same DR antigen were mostly negative or low while those with cells of other groups (carrying different DR antigens; see Table 5) were usually positive. Exceptions to this rule, involving groups D1 and D6, are discussed below.

The right-hand side of Figure 4 shows the serology of the TC groups when the broad or cross-reacting anti-DR sera are considered. Thus, while showing no "cross-reactivity" in MLC, TCs of groups D1 and 2 are all positive for the broad anti-DR (1 + 2) sera. Likewise, cells of D3 and 4 are independent in MLC, but all react with the broad anti-4 sera. The broad anti-5 sera cover the TCs of groups D5, 6, 7, 8, and 10. As indicated, there is a complex relationship between some of the TCs of group D6 and D8 (to be further discussed later), but not among the groups of other TCs which react positively with the broad anti-5 sera. Finally, TCs of groups D9 all reacted with the anti-109 sera which probably "include" antibodies to antigen DR9 (see Figure 1, Table 1, and accompanying text). The association between groups of TCs and the broad or cross-reacting anti-B cell sera as shown in Figure 4 is reminiscent of the situation prevailing in man, where several sets of cross-reacting or supertypic anti-B cell sera (each including some of the established DR antigens) seem to define the separate MB or MT locus.[27,28] For the rhesus monkey, there is as yet no evidence that the broad antisera depicted in Figure 4 define alleles of a *separate* locus of B cell antigens (see also previous sections).

B. The D Locus Antigens

In the late 1970s, several rhesus "harems," each consisting of a few families, were typed with a limited number of TCs in order to study the segregation of the D antigens in families. As expected, parental D antigens segretated together with other RhLA-controlled markers, except in a few recombinant offspring where D and DR antigens were inherited in coupling.[41] Hence, D antigens were probably controlled by the RhLA region, as are A, B, and DR antigens.

In 1980, when 38 TCs defining 10 specificities had become available (Figure 4), the genetic control of those D antigens and their association with DR antigens was further investigated by cellular typing of a population of 93 unrelated animals.[42] The monkeys for this population were randomly selected with regard to their RhLA phenotypes. In

*Two homozygous typing cells of rhesus monkeys derived from offspring of consanguinous matings have been reported by Maurer et al.[40] Unfortunately, those reagents were not available to our team for comparison of the specificities of Maurer's TCs with those defined by our team in Rijswijk.

Table 5
RhLA-CONTROLLED MEMBRANE ANTIGENS OF THE TYPING CELLS

Typing cells			B locus	A locus	DR locus	Ia "other"	48/49
D1	1	i	10/10	17/17	1/1	—	49
	2	i	10/10	17/17	1/1	—	49
	3	f	10	11 29	1	108	48
	4	f	6	11 32	1	108	48
	5	g	10/10	11/26	1/1	—	—
	6	g	6/10	26/26	1/1	—	—
	7	f	6	—	1	103,108	48, 49
D2	8	g	9/33	11/17	2/2	108	48, 49
	9	g	6/19	11/24	2/2	108	48
	10	g	3/5	14/14	2/2	108	—
	11	f	9	11 17	2	108	48
D3	12	g	6/—	13/—	3/3	—	—
	13	g	19/21	24/24	3/3	—	—
	14	g	10/21	13/24	3/3	108	—
	15	g	6/23	24/34	3/3	102	—
	16	f	1	17 26	3 3	—	—
	17	g	6/—	13/17	3/3	—	49
D4	18	g	10/28	2/2	4/4	102	—
D5	19	g	23/23	26/26	5/5	—	—
	20	g	9/23	26/—	5/5	—	—
	21	g	23/—	26/34	5/5	—	—
	22	g	23/—	26/34	5/5	—	—
D6	23	f	6	26	6	108	—
	24	f	19 33	24	6 + 8	108	—
	25	f	6 27	26	6 + 8	108	—
D7	26	g	6/10	20/26	7/7	—	—
D8	27	i	—/—	29/29	8/8	102	—
	28	i	—/—	29/29	8/8	102	—
	29	i	—/—	29/29	8/8	102	—
D9	30	i	—/—	29/29	8/8	102	—
	31	i	6/6	14/14	109/109	—	—
	32	i	19/19	17/17	109/109	—	49
	33	g	9/—	11/34	109/109	—	48
	34	g	9/—	11/34	109/109	—	48
	35	f	6 9	11	109/109	—	—
	36	g	6/19	14/17	109/109	—	48
D10	37	i	10/10	17/17	101/101	103	49
	38	f	10 28	11 26	101	103	—

Note: The letters i, g, and f stand for TCs derived from inbred, genotyped, and feral animals, respectively (see also text). The dash between the two A-, B- and DR- locus antigens for TCs of categories i and g separates the paternal (left) from the maternal (right) contributions to the genotypes. Typing cells 31 to 36 were formerly typed as "blanks" for DR; they show positive cytotoxic reactions with anti-109 sera which are believed to "include" specificity DR9 (see further Figure 1 and accompanying text). Typing cells 37 and 38 show positive cytotoxic reactions with the broad anti-5 sera (see Figure 4) but they also carry 101, which is now considered to be an antigen of the DR series (DR 10; see Figure 1 and accompanying text). For further details regarding the serology, see Roger et al.[12]

From Jonker, M., van Meurs, G., and Balner, H., *Tissue Antigens*, in press. With permission.

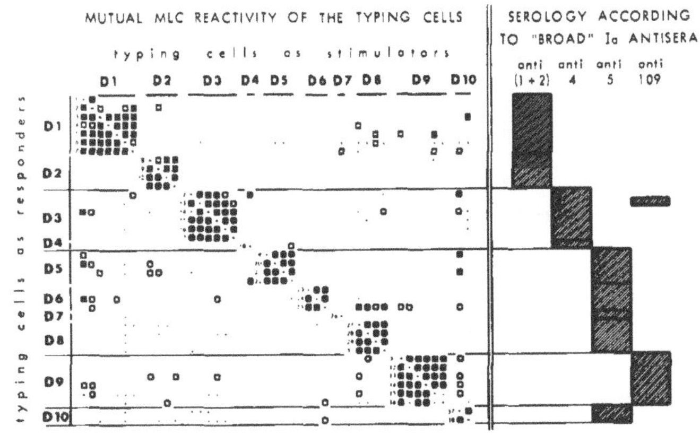

FIGURE 4. Reactivity pattern of 38 homozygous typing cells tested against each other in MLC and serology of the typing cells according to the "broad" Ia-antisera.
Left-hand side of the figure:
■ = negative MLC reaction (median SRR < 50%)
□ = low MLC reaction (median SRR 50 to 60%)
open space = positive MLC reaction (median SRR > 60%)
dot = no data available.
The symbols D1–D10 stand for D locus specificities.
Right-hand side of the figure: The shaded columns indicate the serological phenotype of the 38 TCs depicted on the left, when tested with the broad or cross-reacting Ia antisera in a cytotoxicity test. Thus, TCs of specificity D1 and D2 (cells 1 to 11) reacted positively with B cell antisera of specificity (1 + 2), TCs D3 and D4 with anti-4 sera, etc.

principle, alloimmunized animals were not used*, for reasons to be discussed later. Figure 5 shows the reactivity patterns obtained. Assignment of D antigens was relatively easy when more than two TCs were available, but more difficult and somewhat unreliable when only one or two could be used (D4, D7, D10).

Table 6 again shows the phenotype distribution of the D antigens in the population. Here the calculated gene frequencies** are given together with the observed and expected phenotypes. Since no major differences between observed and expected antigen combinations were found, it is reasonable to assume that we are dealing with allelic antigens of one segregant series which are in Hardy-Weinberg equilibrium. For comparison, the gene frequencies of the serologically defined DR antigens determined independently in a population of 162 monkeys are also given (from Roger et al.;[12] first column of Table 6; in parentheses). It can be seen that the gene frequences for D and DR antigens are of the same order of magnitude, except for antigens DR6 and D6 which are "problematic" in both serological and cellular typing (see below).

*A few animals of the population had a history of alloimmunization but had not been pregnant or intentionally alloimmunized in the year preceding this particular investigation.

**Gene frequencies were determined by using maximum likelihood estimates from the actual phenotype data. Four animals of Figure 5 with possibly more than two D antigens were excluded. Also animals possibly carrying D4, D7, or D10 were excluded from the analysis, since a considerable number of animals had not been tested with the TCs of those specificities.

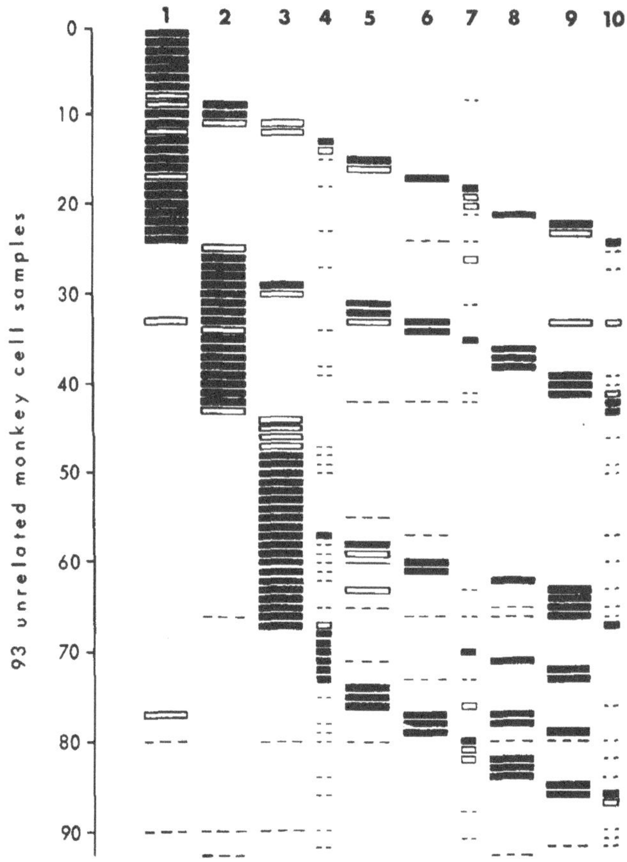

FIGURE 5. Distribution of D locus antigens in a population of 93 unrelated rhesus monkeys.

■ = median SRR value < 50% (typing cell response)
□ = median SRR value 50 to 60% (doubtful typing cell response)
open space = median SRR value > 60% (positive MLC, negative TC response)
--- = no data available

Broad bars indicate that 3 to 7 TCs were available, narrow bars that only 1 or 2 TCs were available to define a specificity. (From Jonker, M., van Meurs, G., and Balner, H., *Tissue Antigens*, in press. With permission.)

C. The Association Between D and DR Locus Antigens

It was clearly of interest to also investigate the correlation between D and DR phenotypes in the population of unrelated monkeys shown in Figure 5.[42] Good correlations (R values of 0.6 or higher) were found between D and DR antigens 1, 2, 3, and 4; weaker ones (R < 0.5) between the D and DR antigens 5, 6, 7, and 8. The correlation between some of the D and DR antigens is illustrated in Figure 6. Monkeys which carried the antigens D1, D2, or D3 are depicted on the left-hand side and almost all carried the corresponding DR 1, 2, or 3 antigens. There were also about 30 animals which carried DR1, 2, or 3 but did *not* type as D1, 2, or 3 (not depicted in the figure); hence, the D phenotypes appear to be "included" in the DR phenotypes. On the right-hand side, similar data are shown for antigens D5, 6, 7, and 8. It is clear that here the D/DR association is not nearly as good as for D/DR antigens 1 to 4. As was the case

Table 6
PHENOTYPE DISTRIBUTION ON CELLS OF 78 UNRELATED RHESUS MONKEYS OF 7 SPECIFICITIES CONTROLLED BY THE D LOCUS

Gene frequency[a]	First antigen	Second antigen	Observed	Expected
0.16	D1	D2	2	3.14
(0.11)		D3	1	3.68
		D5	2	1.79
		D6	1	0.79
		D8	1	1.45
		D9	2	1.78
		D1 or blank[b]	13	9.96
0.13	D2	D3	2	3.00
(0.17)		D5	2	1.46
		D6	1	.64
		D8	3	1.19
		D9	3	1.45
		D2 or blank	6	7.83
0.15	D3	D5	2	1.79
(0.21)		D6	1	0.75
		D8	1	1.40
		D9	1	1.70
		D3 or blank	13	9.44
0.07	D5	D6	0	0.37
(0.08)		D8	0	0.68
		D9	1	0.83
		D5 or blank	4	4.15
0.03	D6	D8	1	0.45
(0.08)		D9	1	0.36
		D6 or blank	0	1.72
0.06	D8	D9	0	0.67
(0.06)		D8 or blank	3	3.30
0.07	D9	D9 or blank	3	4.11
0.33	Blank[b]	blank	8	8.38
$X^2_{21} = 17.64$	$p > 0.50$		78	78.00

Note: At a referee's request, a few changes have since been applied to this table. Thus, while the "message" remains the same, some of the figures in Table 2 of the publication by Jonker et al. differ from those given here.

[a]The gene frequencies of DR antigens determined independently in a different population[12] are given in parenthesis.

[b]The gene frequencies of antigens D4, D7, and D10 were "pooled" here with those of the blanks of the D locus. This was done because of their low frequencies (0.02, 0.02, and 0.01) and because of the limited number of TCs identifying those antigens.

From Jonker, M., van Meurs, G., and Balner, H., *Tissue Antigens*, in press. With permission.

for the typing cells (Figure 4), almost all animals carrying D5, D6, D7, or D8 reacted with the "broad" DR5 sera. The cellularly defined antigens D9 and D10 were found to be correlated with the serologically defined antigens 109 (DR9) and 101 (DR10), respectively, also in the population (see further Table 5, Figure 4, and Jonker et al.[42])

D. Heterogeneity of D Antigens in Relation to B Cell Serology and "PLT"

As can be seen in Figure 4 (upper left corner), the *typing cells of group D1* were not all mutually MLC negative, but appeared to consist of two subgroups. It was earlier postulated that such unidirectional MLC reactivity among TCs of the same D specificity

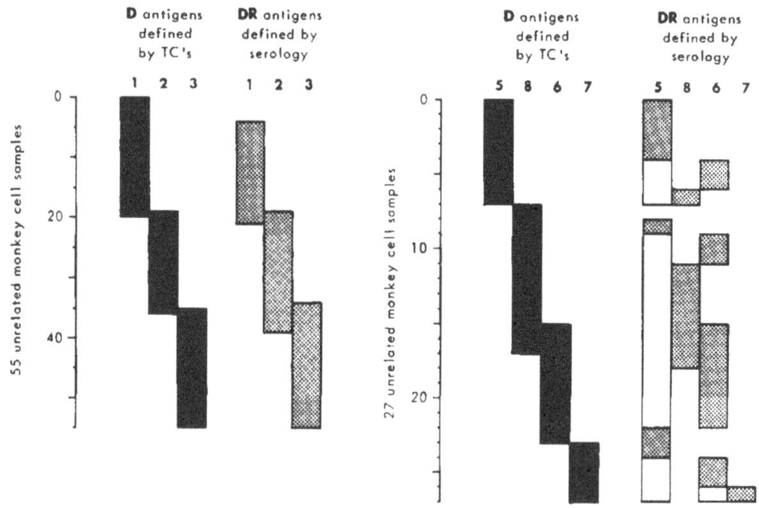

FIGURE 6. Correlation between D and DR antigens in unrelated rhesus monkeys. The data depicted here concern a selection of animals also shown in Figure 5. Animals which carried DR 1, 2, or 3 or DR 5, 6, 7, or 8, but lacked the corresponding D locus antigens, are now shown. The columns on the extreme right of the figure depict the phenotypes for DR 5 to 8. Antigen DR5 is defined by a set of "broad" sera which are cytotoxic also against cells carrying DR antigens 6, 7, and 8. The inclusion is indicated by the boxed-in area of the column for DR5.
■ = presence of a D locus antigen
▦ = presence of a DR locus antigen
(From Jonker, M., van Meurs, G., and Balner, H., *Tissue Antigens*, in press. With permission).

was due to MLC stimulating determinants controlled by another RhLA-linked locus, provisionally called LD2.[34,43,44] However, it was clear from the start that the influence of more than one non-D MLC locus was required to account for all observations. To further investigate this problem, the reaction patterns of cells from the unrelated monkeys of Figure 5 which showed negative or low MLCs with two or more of the TCs of group D1 were analyzed in detail. In the upper part of Figure 7, we depict again the mutual MLC reactivity of the TCs of group D1. In the lower part of the figure, the MLC responses of 22 unrelated monkeys against the 7 TCs of group D1 are depicted. These are essentially the same animals which carried antigen D1 in Figure 5; excluded were two monkeys which themselves provided typing cells and a few others which were not tested with all seven typing cells of group D1. It appears that only a few unrelated responder cells were MLC negative with all seven D1 TCs. Thus, TCs 1 and 2, 3 and 4, and 5 and 6 reacted identically but showed slightly different reactivity patterns, while TC 7 displayed yet another pattern. In this manner, the TCs of group D1 can be subdivided into four "subgroups". On the right-hand side of the figure, the serological B cell phenotypes of the 7 TCs and the 22 unrelated monkeys are given. Interestingly, the TCs of the individual subgroups share (besides DR 1) the same "Ia other" antigens as well as the antigens of the postulated 48/49 series.[26,29,42] On inspecting the B cell phenotypes of the 22 unrelated animals, it appears that the majority does carry DR1 but that there is no correlation between TC responses obtained with any of the "subgroups" of D1 and the serologically defined B cell antigens 108, 48, or 49.

The group of *typing cells defining D6* showed a heterogeneity of a different kind. When tested in matrix fashion with all other TCs (Figure 4), a complex relationship

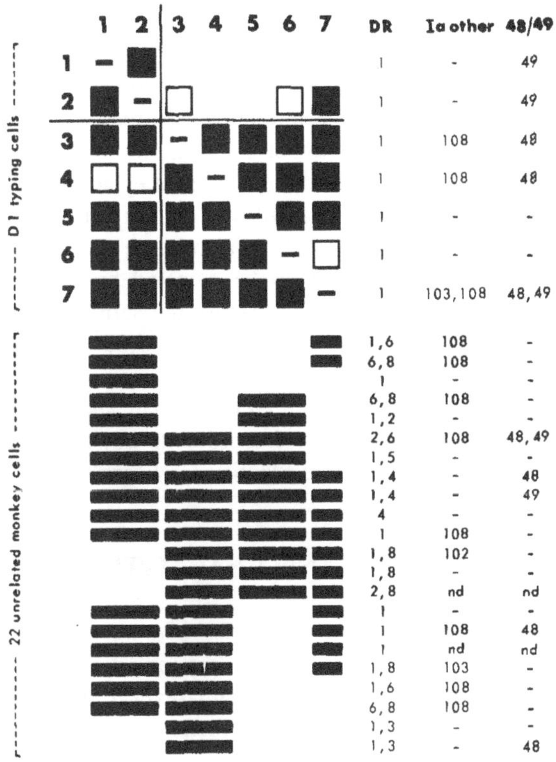

FIGURE 7. Heterogeneity of the D locus specificity D1. The upper part of the Figure shows the mutual MLC responses of the seven TCs of Group D1 as well as their phenotypes for B cell antigens. MLC responses can be negative (black squares; SRR < 50%), questionable (open squares: SRR 50 to 60%), or positive (open spaces; SRR > 60%). The lower part of the figure shows the MLC responses of 22 unrelated monkeys with the seven TCs of group D1 as well as their phenotypes for B cell antigens. Here the black bars indicate a median SRR of <50% (= TC responses), and the open spaces a SRR of > 50% (= questionable or absent TC responses). (From Jonker, M., van Meurs, G., and Balner, H., *Tissue Antigens*, in press. With permission.)

between the TCs of groups D6 and D8 was observed: TC number 25 (D6) was MLC-negative also with the TCs of group D8. Further, the data shown in Table 5 revealed that TCs numbers 24 and 25 of group D6 are both phenotypically DR6 plus DR8! Yet, when the unrelated animals of Figure 5 were tested with TCs of group D6 and D8, they behaved as independent groups of typing cells and the antigens they define did not seem to be "included" in each other. Also, specificities D6 and D8 did not show a positive correlation in an association analysis carried out for the unrelated animals of Figure 5.[42]

A promising approach to analyze the "complexity" of D locus antigens as defined by groups of typing cells (e.g., groups D1 and D6), is the use of "primed lymphocytes" or PLT cells.[45,46] In a conventional PLT test, a lymphocyte sample is primed against a particular D locus antigen and will, usually, show an increased secondary response against lymphocytes carrying the D antigen in question or against "related" antigens. In man and the monkey, the method is being used in attempts to analyze the "complexity'' of D antigens. For this purpose, cells carrying a particular D antigen are sensitized against lymphocytes carrying a possible variant of that same D antigen (or against non-D MLC stimulating antigens; see later). Applying this method, interesting reactivity patterns of PLT cells have been obtained. In man, this approach was systematically used by Mawas et al.,[47] by Shaw et al.,[48] and by Termijtelen;[49] a second MLC locus denoted as SB and linked to HLA could thus be established. In rhesus monkeys, the PLT technique is unfortunately rather difficult to apply and progress in this area has been slow. Yet, preliminary data indicate that, also in the monkey, antigens of a second RhLA-controlled MLC locus are likely to be detected by employing the PLT technique.[50] This second MLC locus is quite possibly the same as has been detected previously by in vivo immunization of rhesus monkeys[51] (see also following sections of the current review).

Thus, the ''complexity'' of D locus antigens in man and the monkey may simply reflect the presence of MLC-stimulating determinants not controlled by the D locus, antigens which apparently cannot yet be identified by serological methods but are detectable by cellular techniques.

V. Ia ANTIGENS AND REACTIVITY IN MLC

A. MLC Loci Other Than D

While there can be no doubt that the D locus antigens are of overriding influence on MLC reactivity in man, the rhesus monkey, and several other species, there is also evidence that antigens *not* controlled by the D locus can stimulate allogeneic lymphocytes in culture (see also above).

In rhesus monkeys, the first evidence for such non-D MLC loci came from mutual MLC testing of the sibs and half-sibs of each harem (see breeding method; Section II). Since the offspring of a harem always share the *paternal* a or b RhLA haplotype, MLC nonreactivity between two individuals (which both carry a or b) implies similarity for the MLC stimulating determinants inherited through the *maternal* RhLA haplotype (from unrelated mothers). The results of systematic testing of offspring within each harem revealed that there must be one or, more likely, several non-D MLC loci also controlled by the RhLA region.[13] Whether these are the analogues of several weak HLA-controlled MLC loci currently proposed also for man[49] is likely, but difficult to prove. As indicated before, the non-D MLC antigens of man and the rhesus monkey cannot be identified by serological methods at present.

A second line of evidence for the existence of non-D MLC antigens of rhesus monkeys came from the observation that a rather large number of RhLA-identical siblings were mutually nonstimulatory in MLC. There were several reasons why those responses could not be the consequence of recombination.[51] Since all the animals displaying aberrant MLC responsiveness had been previously alloimmunized, it was postulated that ''unexpected'' positive MLC responses are likely to be caused or enhanced by alloimmunization or priming. To test that hypothesis, pairs of initially MLC nonresponsive monkeys were intentionally alloimmunized. The first groups were *unrelated monkeys* identical for A, B, and DR locus antigens and mutually MLC negative; after mutual immunization with skin and blood, distinct MLC responsiveness against the partner's lymphocytes occurred in the majority of cases. The second group were MLC-negative *RhLA-identical sibling pairs*. Here, mutual immunization did not lead to significant MLC responsiveness against the partners' cells. However, subsequent immunization with blood and skin from third party animals often led to MLC responsiveness against the original RhLA-identical sibling! The combined results led to the conclusion that: (1) alloimmunization can convert MLC nonresponsiveness into responsiveness due to products of a MHC-linked locus called D′, and (2) the conversion of initially MLC-negative sibling pairs into mutual responders (after immunization against third party tissue) requires the assumption of MLC stimulating determinants which are *not* controlled by the RhLA region (called D″[51]).

It is not yet clear whether the antigens of the D′ locus (loci) are the same MLC stimulating antigens which were described in other parts of this chapter and that alloimmunization simply amplifies the stimulating influence of MLC antigens not controlled by the D locus. Further, it is tempting to speculate that D′ is basically the same locus which can be detected by PLT cells in man and in the monkey. In both instances, alloimmunization (in vivo or in vitro) leads to the detection of MLC stimulating antigens which are controlled by the MHC and which are not yet detectable by current serological methods.* It is also reasonable to assume that the existence of the various non-D MLC loci contributes to the much discussed complexity or heterogeneity of D locus antigens as defined by typing cells homozygous for D antigens (See Section IV).

B. Matching for DR Antigens and MLC Nonreactivity ("Predictive Value" of DR Matching)

Ever since the discovery of the association between D and DR antigens, the question has been posed as to what extent matching outbred individuals (monkey, man) for the readily identifiable DR antigens might predict identity also for D antigens and thus (if we disregard the weaker MLC loci described above) for MLC nonreactivity. The issue of the "predictive value" of matching for DR antigens for MLC nonreactivity became particularly important after it was found that host/donor matching for D and DR antigens significantly improves kidney allograft prognosis in unrelated rhesus monkeys.[51] In this section the relation between D and DR antigens and the predictive value of DR matching for MLC nonresponsiveness will be briefly reviewed. The last section of the chapter will deal with the practical significance of being able to define Ia antigens, i.e., the influence of matching for Ia antigens on allograft prognosis. The question of whether D and DR antigens are identical and controlled by the same gene will not be considered since this issue has been amply reviewed.[25,38] Very briefly, there are many arguments in favor of identity, but also a few rather solid ones against it; the most convincing being the recombinants between D and DR antigens in humans reported in recent years.[53,54] No recombinants between D and DR have yet been found in rhesus monkeys.

For rhesus monkeys, there are also several lines of evidence which support a very close association or identity between D and DR:

1. Homozygous typing cells of a particular specificity are identical for D locus antigens as well as for the associated DR antigens. This holds true for the three categories of TCs described in Section IV. A. of this chapter (consanguinous, genotyped, and feral TCs); the selection of the latter two types of TCs is actually based on that association;
2. An association between D and DR antigens was also found in a random population of monkeys typed for D and for DR antigens (Section IV. C.);
3. The selection of MLC nonresponsive combinations from a population of unrelated individuals virtually depends on prior serological matching for DR antigens.

Yet, in spite of the solid association between D and DR, there are certain observations

*For practical purposes, such as host donor matching for transplantation, it would be clearly of interest to be able to define the non-D MLC stimulating antigens by serology.

which require some comment: (1) the predictive value of DR matching for MLC nonresponsiveness among unrelated individuals does not exceed 40 or 50% of the cases; (2) simultaneous matching for A and B antigens significantly improves the predictive value; and (3) some combinations of DR antigens have a higher predictive value for MLC nonresponsiveness than others.

Tentative explanations for those observations have been reported and can be summarized as follows: first, while D/DR products are incontestably of overriding influence on MLC responsiveness, there are other loci (linked or not linked to the MHC) influencing such responsiveness in monkeys and man. The rather high predictive value of DR matching for MLC nonresponsiveness suggests a relatively weak influence on MLC responses and a low degree of polymorphism for those other MLC loci.[55] Second, additional matching for A and B locus antigens is likely to improve the match also for the products of the other MHC linked MLC loci through linkage disequilibria.[34] Finally, the observation that some combinations of DR antigens have a higher predictive value for MLR than others may be attributable to a better serological definition of some of the DR antigens or, if D and DR antigens are not identical, to a higher degree of association between some of the D/DR combinations in comparison with others.[55]

As indicated earlier, we do not yet have methods for identifying the products of the "other" MLC loci by serological means. Therefore, matching for two DR antigens is still the only practical approach to the selection of unrelated individuals of low or negative mutual MLC responsiveness. The practical significance of DR matching for transplantation immunology is discussed in the following and final section of this chapter.

VI. THE RELEVANCE OF MATCHING FOR D/DR ANTIGENS TO ALLOGRAFT PROGNOSIS

Since man and the rhesus monkey are the only outbred species for which significant information on Ia antigens is available, and since the rhesus monkey is still the most reliable model for preclinical transplantation research, the data accumulated on DR matching and allograft prognosis in the monkey model will be briefly summarized. Where possible, comparisons are made with information obtained (mostly retrospectively) in clinical transplantation.

A. D/DR Matching and the Fate of Kidney Allografts

Ever since the discovery of the human HLA system, the question of whether matching host/donor combinations for the A and B locus antigens of HLA improves kidney allograft survival in man, has been a controversial one. In our opinion, it is of less importance now, since experiments performed in rhesus monkeys have clearly demonstrated that matching for D/DR antigens is of prevailing importance, while identity for A/B antigens does not seem to have a significant influence on kidney graft survival in unrelated rhesus monkeys.[24] Interestingly, if no blood transfusions were given prior to transplantation, the beneficial effect of matching for DR was apparent only in combinations in which the host was MLC nonresponsive against donor lymphocytes.[52] Hence there has been some doubt about whether matching for two DR antigens (regardless of MLC nonresponsiveness) combined with blood transfusions and standard immunosuppression would be the optimal approach toward kidney transplantation.

This clinically important question was again investigated experimentally in the monkey model and the results obtained so far can be summarized as follows. Firstly, DR matching improves kidney graft prognosis in MLC positive as well as in MLC negative combinations, although the effect is most pronounced in the MLC negative combinations; matching for the A and B locus antigens does not seem to have a noticeable influence on the prognosis. Secondly, DR matching and pretransplant blood transfusions have, if anything, an additive effect; again, the effect was most outspoken in the MLC nonresponsive combinations.[56] It is rewarding to note that both these procedures, prospective matching for DR antigens (whenever technically feasible) and the administration of pretransplant blood transfusions, have now become the recommended regimen in the majority of clinical transplant centers.

B. D/DR Matching in Skin Grafting

This issue is of fundamental rather than practical importance, since skin allografting is still a little-used procedure in the clinic (e.g., in cases of severe third degree burns). Previous experiments with monkeys using unrelated host/donor combinations indicated that skin allograft survival was significantly prolonged by matching for A/B antigens.[7] Several years later, when typing and matching for DR antigens had become possible, the positive effect of matching for A/B antigens was confirmed. However, matching for DR antigens did not reveal a noticeable influence on survival times, although a marginal prolongation of graft survival seemed to occur if the host was MLC-nonreactive against the graft donor.[57] Additional data have been obtained during 1980 and 1981. The results confirm that matching for A/B antigens does lead to significant prolongation of skin allograft survival, an effect which is more noticeable in male than in female recipients. Again, DR matching (alone or in combination with matching for A/B antigens) did not have a positive influence on skin graft survival. However, at this time, the beneficial influence on graft survival of MLC-nonresponsiveness of host against donor, was significant. As before, this effect was noticeable only in male recipients and occurred regardless of whether host/donor combinations were matched for A/B antigens or not.

Thus, matching for A/B antigens has a distinct influence on skin allograft survival in unrelated rhesus monkeys and an additional but minor effect of matching for D/DR antigens was observed. Data available for skin graft survival in human volunteers also revealed a positive influence of matching for A/B antigens.[58] The additional positive influence of matching for D/DR antigens seemed more pronounced in man than the effect observed in the monkey model.

C. D/DR Matching in Bone Marrow Transplantation

In clinical bone marrow transplantation, the question of matching *unrelated* host/donor combinations has not been a very important one: the immunological complications occurring in "both directions" (host-vs.-graft and graft-vs.-host reactions) are so severe that, in practice, only HLA-identical siblings have been used as donors of marrow or stem cell concentrates.[59] However, recent developments in gnotobiology (bacterial decontamination) and improved methods of hemopoietic stem cell separation have renewed the interest in the use of marrow from unrelated donors, and thus in the influence of host/donor matching for MHC controlled antigens. And, as has happened before, the monkey model was again employed to provide experimental data which might be extrapolated to man with some degree of confidence.

This is not the place to review the new developments in detail. Suffice it to summarize here some of the monkey data with regard to the influence of matching for A/B and for D/DR antigens. The conclusions were drawn from a variety of experiments using unrelated and related rhesus monkeys, two types of conditioning by total body irradiation, and the use of hemopoietic stem cells as engrafted material.[22] If all available information is considered, the results can be broadly summarized as follows: when an optimum inoculum of hemopoietic stem cells (with regard to the separation method and the number of cells) is transplanted to recipients which have been optimally conditioned by total body irradiation and are optimally supported before and after grafting with regard to infections and hemorrhages, it appears that: (1) matching for A and B locus antigens has a pronounced beneficial influence on the "take" of the graft and on the severity of the Graft-vs.-Host reaction: milder GvH disease, longer survival, better recovery of immune reactivity than in A/B disparate combinations and (2) matching for DR antigens does not have a noticeable beneficial effect on those same parameters, regardless of whether the host/donor combinations are responsive or not responsive in MLC. For further details regarding this subject the reader is referred to the literature.

REFERENCES

1. **Jonker, M., and Balner, H.,** Mixed lymphocyte reactivity in chimpanzees. II. Family studies and identification of D locus antigens, *Tissue Antigens,* 17 (20), 179, 1981.
2. **Lafuse, W. P., Yokota, S., and David, C. S.,** Gene complementations generate hybrid Ia specificities involved immune response, *Transplant. Proc.,* 13 (1), 1017, 1981.
3. **van Vreeswijk, W., Roger, J. H., D'Amaro, J., and Balner, H.,** The major histocompatibility complex of rhesus monkeys RhLA-A. VII. Identification of five new serologically defined antigens, *Tissue Antigens,* 9, 17, 1977.
4. **Balner, H.,** Current knowledge of the histocompatibility complex of rhesus monkeys (A brief review), *Transplant. Rev.,* 15, 50, 1973.
5. **Neefe, J. R., Ellis, E. B., and Rogentine, G. N.,** Rhesus lymphocyte alloantigens, III. Identification of new antigens, *Tissue Antigens,* 6, 195, 1975.
6. **van Leeuwen, A., Schuit, H. R. E., and van Rood, J. J.,** Typing for MLC (LD). II. The selection of non-stimulator cells by MLC inhibition tests using SD identical stimulator cells (MISIS) and fluorescent antibody studies, *Transplant. Proc.,* 5 (4), 1539, 1973.
7. **Balner, H. and van Vreeswijk, W.,** The major histocompatibility complex of rhesus monkeys (RhLA). V. Attempts at serological identification of MLR determinants and postulation of an I region in the RhLA complex, *Transplant. Proc.,* 7 (1), 13, 1975.
8. **Zaalberg, O. B., Fasbender, M. J., and Balner, H.,** Ia-like antigens of rhesus monkeys are distinct from conventional RhLA antigens 2-microglobulin, and immunoglobulin, *Immunogenetics,* 3, 499, 1976.
9. **Widmer, M. B. and Balner, H.,** Inhibition of MLC responder and stimulator function with alloantisera in rhesus monkeys, *Tissue Antigens,* 12 (2), 115, 1978a.
10. **Roger, J. H., van Vreeswijk, W., Dorf, M. E., and Balner, H.,** The major histocompatibility complex of rhesus monkeys. VI. Serology and genetics of Ia-like antigens, *Tissue Antigens,* 8 (1), 67, 1976.
11. **Roger, J. H., van Vreeswijk, W., D'Amaro, J., and Balner, H.,** The major histocompatibility complex of rhesus monkeys. IX. Current concepts of serology and genetics of Ia antigens, *Tissue Antigens,* 11, 163, 1978.
12. **Roger, J. H., van Vreeswijk, W., and Balner, H.,** The major histocompatibility complex of rhesus monkeys. XIII. Current knowledge of DR and other B-cell specific antigens, *J. Immunogenet.,* 7, 333, 1980.
13. **Balner, H. and Toth, E. K.,** The histocompatibility complex of rhesus monkeys. II. A major locus controlling reactivity in mixed lymphocyte cultures, *Tissue Antigens,* 3 (4), 273, 1973.
14. **Bradley, B. A., Edwards, J. M., Dunn, D. C., and Calne, R. Y.,** Quantification of mixed lymphocyte reaction by gene phenomenon, *Nature (New Biol.),* 240, 54, 1972.
15. **Mempel, W., Grosse-Wilde, H., Baumann, B., Netzel, B., Steinbauer-Rosenthal, I., Scholz, S., Bertrams, J., and Albert, E. D.,** Population genetics of the MLC response: typing for MLC determinants using homozygous and heterozygous reference cells, *Transplant. Proc.,* 5, 1529, 1973.
16. **Ryder, L. P., Thomse, M., Platz, P., and Svejgaard, A.,** Data reduction in LD-typing, *Histocompatibility Testing,* Munksgaard, Copenhagen, 1975, 557.
17. **van Rood, J. J. and van Leeuwen, A.,** Leukocyte grouping. A method and its application, *J. Clin. Invest.,* 42, 1382, 1963.

18. **Gabb, B. W., Piazza, A., D'Amaro, J., and Balner, H.,** Genetics of the RhL-A system of rhesus monkeys, *Transplant. Proc.* 4 (1), 11, 1972.
19. **Balner, H., Cleton, F. J., and Eernisse, J. G.,** Eds., *Histocompatibility Testing 1975*, Munksgaard, Copenhagen, 1975.
20. **Marquet, R. L., Heystek, G. A., and van Bekkum, D. W.,** Heterotopic heart and kidney transplantation in rhesus monkeys, in *Medical Primatology, Part II*, Goldsmith, E. I. and Moor-Jankowski, J., Eds., S. Karger, Basel, 1972, 125.
21. **Balner, H.,** Skin grafting in monkeys and apes, *Transplantation*, 8 (2), 206, 1969.
22. **Wagemaker, G., Vriesendorp, H. M., and van Bekkum, D. W.,** Successful bone marrow transplantation across major histocompatibility barriers in rhesus monkeys, *Transplant. Proc.*, 13, 875, 1981.
23. **Albrechtsen, D., Solheim, B. C., and Thorsby, E.,** Serological indefinition of five HLA-D associated (Ia-like) determinants, *Tissue Antigens*, 9 (3), 153, 1977.
24. **van Es, A. A., Marquet, R. L., van Vreeswijk, W., Tank, B., and Balner, H.,** The influence of matching for RhL-A (SD) antigens and of mixed lymphocyte reactivity on allograft survival in unrelated rhesus monkeys, *Transplant. Proc.*, 9 (1), 257, 1977.
25. **Bodmer, J. G. and Bodmer, W. F.,** Summary and conclusions, in *Histocompatibility Testing 1977*, Bodmer, W. F. et al., Eds., Munksgaard, Copenhagen, 1978, 351.
26. **Balner, H., van Vreeswijk, W., and Roger, J. H.,** The major histocompatibility complex of rhesus monkeys. XIV. A family study of DR and other RhLA-linked cell membrane antigens, *Tissue Antigens*, 17, 530, 1981.
27. **Duquesnoy, R. J., Marrari, M., and Annen, K.,** Identification of an HLA-DR associated system of B cell alloantigens, *Transplant. Proc.*, 11, 1757, 1979.
28. **Park, M. S., Terasaki, P. I., Nakata, S., and Aoki, D.,** Supertypic DR groups: MT1, MT2 and MT3, in *Histocompatibility Testing 1980*, Terasaki, P. I., Ed., UCLA Tissue Typing Lab., Los Angeles, Calif., 854.
29. **Jonker, M., van Meurs, G., and Balner, H.,** Typing for RhLA-D in rhesus monkeys. I. Characteristics of ten groups of homozygous typing cells, *Tissue Antigens*, 1982b, in press.
30. **Giphart, M. J.,** personal communication, 1981.
31. **Roger, J. H., van Vreeswijk, W., and Balner, H.,** 1982, manuscript in preparation.
32. **Meo, T., David, C. S., Rijnbeek, A. M. Nabholz, M., Miggiano, V. C., and Shreffler, D. C.,** Inhibition of mouse MLR by anti-Ia sera, *Transplant. Proc.*, 7 (1), 127, 1975.
33. **Giphart, M. J., Tank, B., Bruning, J. W., and Balner, H.,** The major histocompatibility complex of rhesus monkeys. VIII. Isolation and partial characterization of SD antigens, *Transplantation*, 25, 131, 1978.
34. **van Es. A. A. and Balner, H.,** The RhLA complex of rhesus monkeys. X. Implication of the association between D and Ia_1 locus antigens, *Tissue Antigens*, 12 (4), 279, 1978a.
35. **Ziegler, J. B., Alper, Ch. A. and Balner, H.,** Properdin factor B and histocompatibility loci linked in the rhesus monkey, *Nature (London)*, 254 (5501), 609, 1975.
36. **Dorf, M. E., Balner, H., and Benacerraf, B.,** Mapping of the immune response genes in the major histocompatibility complex of the rhesus monkey, *J. Exp. Med.*, 142, 673,1975.
37. **Balner, H., Dorf, M. E., de Groot, M. L., and Benacerraf, B.,** The histocompatibility complex of rhesus monkeys. III. Evidence for a major MLR locus and histocompatibility linked Ir genes. *Transplant. Proc.*, 5 (4), 1555, 1973.
38. **Balner, H.,** Are D and DR antigens identical? A review of available data for man and the rhesus monkey, *Transplant. Proc.*, 11 (1), 657, 1979.
39. **Khan, P. M., Volkers, P. W. S., Doppert, B. A., Bijnen, A. B., Schreuder, I., and van Rood, J. J.,** The locus for Glyoxylase I (GLO) is between HLA-A and PGM3 on chromosome 6 of man (Baltimore Conference, 1975, 3rd International Workshop on human mapping), *Births Defects: Original Article Series*, 12 (7), 328.
40. **Maurer, B. A., Siwarski, D. F., and Neefe, J. R.,** Definition of two LD antigens in rhesus monkeys, *Tissue Antigens*, 13, 81, 1979.
41. **van Es, A. A. and Balner, H.,** The major histocompatibility complex of rhesus monkeys. XII. Cellular typing for D locus antigens in families, *Tissue Antigens*, 13, 255, 1979a.
42. **Jonker, M., van Meurs, G., and Balner, H.,** Typing for RhLA-D in rhesus monkeys. II. Genetics of the D antigens and their association with DR antigens, *Tissue Antigens*, 1982a, in press.
43. **Widmer, M. B. and Balner, H.,** Multi-locus control of MLC reactions in rhesus monkeys, *Immunogenetics*, 7, 155, 1978b.
44. **Jonker, M. and Balner, H.,** Current knowledge of the D/DR region of the major histocompatibility complex of rhesus monkeys and chimpanzees, *Human Immunol.* 1 (4), 305, 1980.

45. **Sheehy, M. J., Sondel, P. M., Bach, M. L., Wank, R., and Bach, F. H.,** HLA-D (lymphocyte defined) typing: a rapid assay with primed lymphocytes, *Science*, 108, 1308, 1975.
46. **Termijtelen, A., Bradley, B. A., and van Rood, J. J.,** A new determinant, defined by PLT, coded for in the HLA region and apparently independent of the HLA-D and DR loci, *Tissue Antigens*, 15, 267, 1980.
47. **Mawas, C., Charmot, D., Sivy, M., Mercier, P., North, M. L., and Hauptmann, G.,** A weak human MLR locus mapping at the right of a crossing-over between HLA-D, Bf and GLO, *J. Immunogenet.*, 5, 383, 1978.
48. **Shaw, S., Johnson, A. H., and Shearer, G. M.,** Evidence for a new segregant series of B cell antigens which are encoded in the HLA-D region and stimulate secondary allogeneic proliferative and cytotoxic responses, *J. Exp. Med.*, 152, 565, 1980.
49. **Termijtelen, A.,** HLA-Linked Alloactivating Determinants, Ph.D. thesis, University Leiden, Pasman, The Hague, 1981.
50. **Jonker, M. and Saltzman, S.,** The use of the PLT test to investigate the "complexity" of D antigens in rhesus monkeys, Annual Report REP Institutes 1979, 339.
51. **van Es, A. A. and Balner, H.,** The RhLA complex of rhesus monkeys. XI. Positive MLC reactions between identical sibs caused by alloimmunization, *Tissue Antigens*, 12, 239, 1979b.
52. **van Es, A. A. and Balner, H.,** Serological matching for D-locus antigens improves kidney allograft survival in rhesus monkeys, *Transplantation*, 26 (3), 187, 1978b.
53. **Suciu-Foca, N., Godfrey, C., Rohowsky, C., Khan, R., Susinno, E., and Hardy, M.,** HLA-D-DR relationships. V. A cross-over between HLA-D and DR, in Histocompatibility Testing 1980, Terasaki, P. I., Ed., UCLA Tissue Typing Laboratory, Los Angeles, 1980, 881.
54. **Sachs, J. A., Jaraquemada, D., and Festenstein, H.,** Intra HLA-D region recombinant maps HLA-DR between HLA-B and HLA-D, *Tissue Antigens*, 17, 43, 1981.
55. **Borleffs, J. C. C., de By-Aghai, Z., and Balner, H.,** Variable "predicitive value" of DR matching for MLC nonresponsiveness in rhesus monkeys, *Transplant. Proc.*, 13 (1), 1057, 1981a.
56. **Borleffs, J. C. C., Marquet, R. L., and Balner, H.,** Pretransplant bood transfusions have an additive positive effect on kidney graft prognosis in D/DR matched rhesus monkeys, *Transplantation*, 32(1), 48, 1981b.
57. **van Vreeswijk, W. and Balner, H.,** MHC-matching and other factors influencing skin allograft survival in related and unrelated rhesus monkeys, *Transplantation*, 30 (3), 196, 1980.
58. **Jonker, M., Hoogeboom, J., van Leeuwen, A., Koch, C. T., Blussé van Oud Alblas, D., and van Rood, J. J.,** Influence of matching for HLA-DR antigens on skin graft survival, *Transplantation*, 27 (2), 91, 1979.
59. **van Bekkum, D. W.,** Summing up in perspectives, in *Immunobiology of Bone Marrow Transplantation*, Thierfelder, S., Rodt, H., and Kolb, H. J., Eds., Springer-Verlag, Berlin, 419, 1980.

INDEX

D

E

F

G

H

I

J

K

L

M